From the Books of

CHRISTY MITCHELL

Pebbles as a Hobby

Pebbles as a Hobby

JANET BARBER

Hippocrene Books, Inc—New York

First published in Great Britain by Pelham Books Ltd.

Hippocrene Books, Inc.
171 Madison Avenue
New York, N.Y. 10016

First published in the United States 1972

Library of Congress Catalog Number 72–86851

ISBN 0–88254–026–2

Printed in Great Britain

To Sally Adams

Contents

Illustrations

COLOUR PLATES

BLACK-AND-WHITE PLATES

LINE DRAWINGS

Drawings by Nicholas Wegner
Photographs: All colour plates by Colin Foo.;
Black and white Plates 1 and 6 Robilt Ltd.;
Plates 2–5 by Fritz Wegner;
Plate 7 Colin Foo;
Plates 8 and 9 Expo Tools & Appliances Ltd.

Introduction

This book is intended for all those people who have ever picked up a pebble and wondered what it was. It attempts to give information about pebbles in simple language, so that anyone who is daunted by the technicalities of geology and mineralogy will be able to identify a wide range. This includes methods for carrying out home tests, in conjunction with identification charts.

Although every country has its own geology, and you may live where there is a different range of rocks and minerals, most of those in this book are available all over the world—in fact, all but one exist in other countries. There may, for example, be differences in scale between agate found in Britain and agate found elsewhere, and there may be agate variations elsewhere not available here, but the composition and form will be basically the same. And, for those who want to take the pebble hobby further than collecting, advice is given on using tumbler polishers and lapidary machines, as well as ideas for things to do with pebbles.

I should like to acknowledge gratefully the help and information I have been given by many people. To Barbara Hargreaves who suggested this book, and who, among many friends, lent me pebbles and jewellery to photograph. In this context I should like to thank Mr. C. O. George, managing director of Ammonite Ltd. who parted with a number of his own collection of beautiful polished pebbles, and was extremely tolerant in leaving them with me for several months. Some are shown in the book jacket.

'Rock shops' are fairly new in Britain, although they have been long established in other countries. This is the nickname for shops dealing in gems and minerals, lapidary and sometimes geology equipment. They tend to be a source of information on the subject to many customers, who come in carrying boxes and packages from which they produce mineral and pebble specimens to be identified, or at least talked about. Such requests are usually dealt with most helpfully by the patient rock shop traders, and as the other customers are known to join in the discussion sometimes, it gives rock shops an unusually sociable atmosphere.

I myself, while preparing this book, have been given helpful information by several people, and my thanks go to Mr. Rowley Collier, to Mr. Pistone of Craftorama Ltd., London; also to Mr. Brian More; and to Mrs. M. L. Beach, of the firm of the same name in Twickenham, Middlesex.

I should also like to thank Mr. Sidney Bendall, N.D.D., F.S.D.C., designer-craftsman, for his advice about working with stones and making mosaics, and Mr. Keith Mossman, author and gardening expert, for his advice about making pebble gardens.

My thanks go, too, to Mrs. Yvonne Drinkwater, Mr. John Hall, David Stone and Adam Brien, and to my daughter Elizabeth, for their kind help.

A person who works on beaches all the year round tells me that, after a fierce storm with high winds and raging seas, a beach and all its larger pebbles can disappear overnight, lost to the force of the sea. The next day, given a wind from the right direction, and a powerful longshore drift, fresh piles of shingle and pebbles are heaped along the foreshore. The beach is constantly changing—being added to and taken away by the action of wind and water.

It is this image of the sea and nature which makes pebbles, for me, such a potently captivating hobby.

Janet Barber

CHAPTER ONE

The Pebble Hobby

Pebbles? You may wonder what on earth anyone can do with these apparently commonplace rounded stones as a hobby more than bringing the odd attractive specimen home from the beach, and placing it on the mantelpiece as a holiday souvenir.

The answer is you can do many different things with pebbles, from collecting them to polishing them and making them into jewellery or ornaments and pictures; from painting them with patterns and designs, to using them decoratively in or out of doors.

A drift of shingle along the foreshore gives the impression of being all brown and black, with an occasional white shell or chalk pebble. Rocks weathered on the sand have a generally faded appearance, an occasional glint of pyrites or mica revealing more interesting things to discover. But pebbles are, in fact, as varied as wild flowers and animals, and far from being in muted colours, they exist in all the hues of the paintbox, all the luminous tints of the rainbow, and vary in texture from granular or cindery rock to clear, transparent, hard crystal.

You find them on the beach, or wherever there is running water—fragments broken from massive rocks by the action of the elements, eroded by underground streams, the roots of plants and ice in winter. They have been slowly formed by the action of water moving them along, and tumbling them about, knocking off their corners and wearing them smooth, depending on the hardness of the original rock.

Pebble collecting includes treasure hunting. The semi-precious minerals chalcedony, agate, amethyst can be found, with luck, on the beach, as water-worn fragments or pebbles. Jasper and quartz are not uncommon, and these stones have been used throughout history for jewellery and objets d'art.

Flint, by no means precious, but plentiful on south-east beaches, is made of the same mineral as agates and quartz. This may surprise you at first—but it is the sort of thing that you learn when you collect a couple of handfuls of pebbles and take time to identify them by certain characteristics which do not require any special training, knowledge or scientific equipment to observe. You can even learn to recognise gem crystals embedded in rocks. They do occur in certain rocks in this country, though they are usually small. Garnet and tourmaline are two kinds you could hope to find.

Fossils and shells are bound to come your way if you go in for the pebble hobby. Some rocks are a mass of little shells cemented together. Others have fossils embedded in them like perfect carvings or raised replicas of snail-shells, millions of years old sea creatures in stone. This is another side interest of the hobby.

From your interest in pebbles, inevitably and painlessly you will learn something about geology, the science of the earth's crust. At its most serious geology can seem a daunting study, but it is a much more exciting subject than some might imagine. You will be bound to find out a little about mineralogy, the science of the chemicals of which rocks are made, through collecting them in their beautiful crystal forms, and a fraction about crystallography, at the least, which is the study of the shapes and forms—pointed and flat, cubes and chevrons—of glittering mineral crystals.

It is almost impossible to travel anywhere in the world by land without picking up pieces of broken rock, dolomite from the Dolomite Mountains in the Alps, conglomerate in the U.S.A., rocks which you find in Britain too. In the desert you could look for wind-sculpted pebbles called 'driekanters'; in Scandinavia, there is porphyry, rock containing large crystals. There is porphyry in Britain too.

It is clear from this that rocks, pebbles take you out of doors, onto the beach, into the country, picnicking, and even camping, in order to stay and search in a remote area where there are specimens you particularly want. It can reach a point where one of the things you look up about a place that you are going to visit is the local bedrock! On a geological map, of course.

Not all of us may have the energy and opportunity to go and find our own pebbles and rocks, nor is it essential to be able to do so. You can buy everything you need from a rock shop, if you prefer. These shops have grown up with the flourishing 'rock hobby' in the United States, where they have existed for a number of years. Lately more and more have opened in Britain, and the leading firms operate extensive mail order services.

In these rock shops you will find a wide selection of polishing machines and amateur lapidary equipment, as well as the stones themselves. They sell grits and polishing powders for polishing pebbles by hand and tumble polishing, and small, as well as large lapidary machines, suitable for schools and clubs, which will deal with commercial quantities of stone.

Some shops stock beach pebbles, but more often the stones are rough broken pieces of imported semi-precious gem rock, pale green amazonite, transparent rose quartz, and shot-silk tigereye, a brown-and-gold striped stone from Africa.

Taking the hobby indoors, its wet weather, dark winter evening side is perhaps its most popular side. There is a wide range of things to do then, with pebbles. You can use shop-polished stones and bought jewellery mounts, ring blanks, brooch settings, cuff links etc., to make up into custom jewellery eminently suitable for your Christmas and birthday presents, bazaar contributions, and the like.

Better still, you can polish your own pebbles, especially if you own a tumbler polisher, which is a machine which tumbles stones in water in a barrel, with the addition of grits and polishing agents. These machines are very simple to use, and there are many inexpensively-priced hobby models on the market.

It is really amazing what marvellous results you can obtain by grinding off the dull weathered surface of an ordinary pebble picked up out of doors and polishing it. The polishing reveals intense colours and sometimes beautiful patterns, and like the ugly duckling who turned into a swan, the pebble is suddenly seen as a natural work of art.

If you go far enough into working with stones, you may even possibly become an amateur gemmologist, learning to facet semi-precious stones, although by the time you have become as involved as this you will almost certainly want to go to an evening class in the subject, or join a lapidary society to extend your skill with expert help.

The equipment you can use can amount to a lot, or a little. By joining a class or a club, you can probably share in the machines provided, until you decide you would like your own at home. This is perhaps a good way of beginning, because when you know more about the hobby in practice, and have used the equipment, you will have a more informed idea of what is involved, and what you will need.

Expense is bound to be a concern, but the running costs of the hobby are small, especially if you find your own stones out of doors. Once you have whatever equipment you have decided to buy initially, or have been given, costs are likely to be mainly for grit and polishing powder, abrasive paper, and electricity to run the motor, with perhaps bought rough gem rocks if you want a particular kind from overseas, and occasional optional investments in accessories for your equipment.

If you don't want to buy machines, you can obtain slabbed as well as polished stones from the rock shop, and some firms will saw rock or trim it to size for a fee. Firms generally advertise in lapidary magazines.

Small motorised tumbler polishers are available from about £5. For average individual use, a machine which takes a 3 lb. barrel would be adequate, and a second barrel would be essential for use at the polishing stage. Including grits, and polishing agents, you

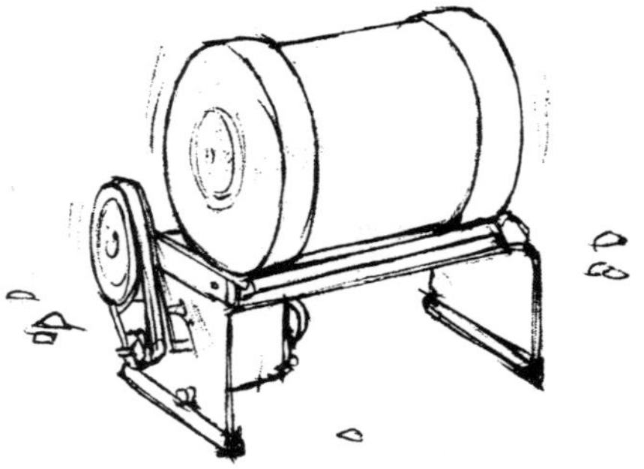

FIG. 1 A simple tumbler polisher

could expect to spend from about £7 to £20 at the outset, more or less. A simple model without such refinements as three-speed controls and superior barrels does a good job, provided that it is a branded make, but as with motor cars and other equipment, it is sometimes worth paying a little more to have better-quality materials and a few refinements. There are, of course, tumble polishers costing vastly more than these figures, but they would be for the dedicated enthusiast or for commercial use.

Lapidary equipment can range from hand-polishing materials for less than £1 to electrically-powered, all-purpose lapidary machines on which, by changing the accessories, you can saw, grind, shape and polish pebbles and pieces of rock. At present, there is at least one really junior-sized lapidary machine on the British market which costs under £10 for the basic, unmotorised lapidary unit (the motor costs extra, from £5 upwards), which at the time of writing you can fit up with grinding and sanding wheels, and a diamond saw blade for about £20, excluding the price of the motor.

This is a very small machine. You can see what others are available by sending round to the leading rock firms, and asking for copies of their catalogues. At present you can obtain excellent lapidary machines with accessories from about £30, plus the price of the motor. It is of course possible to pay hundreds of pounds for precision-built, professional lapidary equipment, which is suitable for commercial or group use.

One firm, Kernocraft Rocks and Gems Ltd., 44 Lemon Street, Truro, Cornwall, supplies kits for making tumbler polishers, and parts for building a lapidary machine. Details and prices are given in the firm's illustrated catalogue, and an important point, it also supplies competitively priced second-hand motors. These would enable you to build you own machines, although you would still need to buy grinding wheels and diamond saw blades.

Each of the lapidary processes is catered for by individual machines, which do offer definite advantages in many ways, since each one concentrates on a particular aspect of working in stone. They can offer a less cramped layout and, on most of them, the boring process of having to change the accessories for each job is cut out.

To have a separate machine for every process, however, would add enormously to the outlay, running expenses and space needed for the hobby. If all the same, you are keen to do this, it is suggested that your basic workshop should consist of a tumbler polisher, which should be a strong, simple efficient machine with well-constructed barrels; a trim saw, or trim and slab saw, and a lap wheel, which is a flat, power-driven wheel, with cast-iron and other laps, used for grinding and polishing rock faces. Other machines could be acquired later, and as already said, there is a wide range of combination machines which will perform all these processes efficiently for the amateur hobbyist or minor workshop.

This, then, is the pebble hobby. An interest for summer and winter, indoors and outdoors. Scientifically fascinating, and creatively satisfying. In the following chapters you can find information about the equipment you need for outdoor trips collecting pebbles, as well as lapidary equipment. Charts give descriptions of over seventy kinds of pebble, so you can identify yours, as well as some of gem crystals and fossils and shells. How to saw and grind and polish your pebbles comes in the later part of the book, and you may find some of the ideas for things to make in the last section not only helpful, but a concealed way of learning some of the decorative techniques used with stones.

Useful addresses are given at the end of the book, and wherever possible the text is illustrated to add to your information.

CHAPTER TWO

The Pebbles

In order to be able to name your pebble, to tell the difference between one kind and another, it helps to know where it came from, and how it was formed. As already said, pebbles are fragments which have broken off larger masses of rock as a result of rain and water accumulations seeping into the cracks and crevices. These fragments are caught up in streams or strongly flowing rivers, and tumbled about, knocking into each other and rocks on the river bed, wearing their rough surfaces smooth in the process. Ice, changes of temperature from hot noon to freezing midnight, all help to erode rock and, on the surface, frost, wet, tree roots and moss all do their damage.

Except where there are mud flats, you are likely to find plenty of pebbles by the sea. Small granite pebbles among the boulders by the Atlantic; whole beaches of shingle near chalk cliffs, consisting of hundreds of thousands of flint pebbles and others, with shells and seaweed and bottle glass and brick among them.

These pebbles have been formed by the action of the sea endlessly breaking against the land, undercutting cliffs. The coast has to withstand storms hurling tons of water to smash against the shore, sending up spray as high as houses. Even on calm days the endless rhythm of the waves breaking and receding wears the pebbles.

The wind blows waves into the shore at an angle, called the 'longshore drift' and recedes, the 'backwash', straight down the beach, dragging the shingle with it with a swishing sound. The pebbles are moved with the drift and tumbled over each other as they travel, sometimes, barring obstacles, some way along the coast. This is the reason why you occasionally find specimens from far away, among the prevalent local rocks.

Pebbles can be ice-worn, seared by glacier conditions, or wind-worn, formed in the path of a strong prevailing wind. Wind-worn 'driekanters' are three-edged pebbles, and are peculiarly beautiful, because sand grains blown across them steadily and abrasively have sculpted them into sheared-off faces with edges, as if they had been gracefully cut off by the wind.

The shape of the pebble will be partly due to the structure of the rock it came from. A flat pebble probably came from a layered rock like shale; a solid humpy stone with barely smoothed irregularities is probably from some tough, durable rock, such as granite. If you have found a pebble with bands of white crystals standing out in ridges from the rest, you may have a sandstone pebble with quartz veins, and the sandstone has worn away faster than the quartz.

Rocks are usually a mixture of several minerals, although some, like chalk, consist largely of a single mineral. A mineral has a chemical composition as well as other characteristics, and minerals which are often found in rocks, such as feldspar or olivine, are called rock-forming minerals.

The age of rocks can be counted in hundreds of millions of years, and even the youngest geological era is given a million years. Rocks are divided into three main groups, igneous, sedimentary and metamorphic. If you know something about each of these groups, you will be able to pick out the characteristics of rocks from them more easily, which will help you in identifying your pebbles.

Igneous rocks have formed from molten material which cooled slowly, deep down in the earth, and hardened into tough, granular coarse textured rock; or, when erupted onto the earth's surface and cooling rapidly, as with volcanic rock, solidified into fine-textured, even glassy rock.

To see an example of igneous rock which formed below the earth's surface, you could look at some unpolished granite. In granite you can see the mineral components quite clearly, pink, grey or white feldspar, clear quartz crystals, and black mica specks. On the other hand, volcanic rock is smooth and, to the eye, non-crystalline—although, under the micrscope, its tiny crystals can be seen.

If you have ever visited Fingal's Cave in the Isle of Staffa off the south-west coast of Scotland, and have seen the amazing columnar formation of the rocks there, then you have seen volcanic rock. In this case, it was basalt. Not surprisingly, volcanic rocks look like material which has been subjected to fiery heat. Lava can be slaggy or cindery. Tuff is volcanic ash; obsidian, black volcanic glass, and is amorphous, that is, non-crystalline, and completely smooth in texture.

Sedimentary rocks, the largest group of rocks, are formed from sediments of mud and sand, and debris of shells and skeletons settling and compacting under pressure from new sediments above. Chalk, limestone, sandstone are in this group. One attractive feature of sedimentary rocks is that you will find fossils and embedded shells in them. These fragile remnants of past living things would never survive the violent heat and pressure forming igneous rocks, or the other group, metamorphic rocks.

Metamorphic rocks are rocks which have been altered, and their mineral materials reconstituted from sedimentary or igneous rocks. Marble is metamorphic limestone; slate is metamorphic shale. 'Metamorphic' a geological term, is defined in the Oxford Dictionary as 'rock that has undergone transformation by natural agencies'. Igneous and sedimentary rocks, which have been subjected to huge pressure by folding and upheaving movements in the earth's crust, are heavily compressed, into layers, or, if banded, are distorted so that you can see the folding and deformed banding, or the minerals are reformed as in marble or metamorphic granite called 'gneiss'.

If you have a pebble which reminds you of granite, with the same hard granular texture, but with the black specks of mica in horizontal bands, this is probably 'gneiss' (pronounced 'nice'). You can see the pressure lines in the laminations in slate, or the layers like thin shining pages or leaves in a glassy book of mica. When identifying your pebbles against the charts, you will find it helpful in classifying them if you remember these characteristics.

Mineral pebbles, which include the semi-precious gem pebbles, are not so plentiful, being as a rule secondary formations in rock. These pebbles include quartz, a very hard crystalline mineral which occurs in cavities in igneous rocks, and also clear rock crystal, mauve amethyst, yellow citrine, smoky cairngorm, and various other types.

Agates, which are banded pebbles, belong to a group called 'chalcedony', which forms in lava cavities. You can often pick up fragments of blue-grey chalcedony on the beach if you know what to look for. There are several kinds of agate, usually distinguished by the pattern of their banding, and pebbles of the chalcedony group have different names depending on the colour or other features; carnelian, for example is flesh-coloured chalcedony, moss agate is chalcedony with chemical inclusions which look like green plants petrified inside the mineral.

Flint is a common mineral, both as pebbles, and as nodules formed in chalk seams and cavities. Chert, which is almost identical to flint, occurs in limestone.

All the mineral pebbles named here are made of the same material, the mineral silica. Although pebbles are often moved far from their place of origin by water, or by being carried by man for building purposes or ballast, you can hope to have a good chance of finding specimens in areas of the rock in which they form.

Quartz formed in igneous rocks is crystalline, with something of a sugary texture when seen under the magnifying glass (although its crystals are a different shape from those of sugar); chalcedony, like the volcanic rocks in which it forms, is crypto-crystalline, that is, it appears to be as smooth as honey. Flint and chert are crypto-crystalline, or smooth-textured, too.

There are other mineral formations which are bound to captivate the magpie collector—formations which occur in massive amounts in specific sites. Brittle, lovely crystals of fluorspar, for example, and show-white gypsum alabaster. Many crystals are too soft or fragile to stand up to the pounding it takes to make a water-worn pebble, but they can be entrancingly beautiful, like branched diadems, or large, faceted jewels, sometimes brilliantly tinted or phosphorescent.

These are only some of the many different kinds of pebbles. Metallic crystals, gold-coloured pyrites, turn up on the beach sometimes. Amber, which is lurid, orange-yellow-coloured fossilised resin, can occasionally be picked up on east coast beaches, washed in from the North Sea. Jet, another type of pebble, is intensely black, lightweight fossilised wood, which used to be fashionable in Victorian times. In England it is found near Whitby in Yorkshire.

WHERE TO LOOK FOR PEBBLES

The main place for pebbles is the beach, but wherever there is a fast moving watercourse you can hope to find some. It is a casual search in the beginning. Pebbles are easily found and collected. But as you become more experienced, you will begin to hanker after trips to areas where the pebbles are different from those nearer home, and perhaps you will visit famous beaches like the Chesil Beach in Dorset, and start persuading willing relatives to cart back heavy bags of rocks from their travels!

There are opportunities for finding pebbles inland where there are rapid tumbling rivers and streams, as in mountainous areas. Lakes and shallow brooks are often too muddy at the edges to be very rewarding to the pebble hunter, unless they are in wilder, more rocky barren places.

Quarries and cuttings into the land, mines and industrial excavations expose buried seams and layers of rock, and are favourite hunting grounds for collectors looking for rock and mineral specimens. It is very important to bear in mind, if you should decide to look for specimens in these places, that apart from the need to have permission to go into them, they can be extremely dangerous, and great care has to be taken when exploring or climbing about in them. A note about safety is given in chapter 3.

When you realise the fact that water-worn pebbles are made of rock, or mineral solids formed in rock, and that, essentially, it is the rock itself or the mineral which is the prime interest of the pebble, then the whole country becomes your hunting ground.

If you wish that you could go in quest of pebbles more purposefully, and that you could have a map which would tell you precisely where a certain kind of rock or mineral is to be found, there is an answer. There are maps which give this information, called geological maps; they are published by the Institute of Geological Sciences, and on sale at the Geological Museum, Exhibition Road, London, S.W.7.

The snag is, that for most of us, geological maps are not as simple to read as road maps. They show not only the solid 'bedrock' of the area, if they are what are called 'solid' maps, but also the drift of mud and sedimentary deposits over the surface, often many feet deep, if they are what is called 'drift' or 'drift and solid' maps.

Geological maps show the rock features of an area in colour, or in black and white patterns, with various symbols to tie in with a key. Although some of the rocks in the key are familiar names like 'limestone' or 'shales', others make very mystifying reading to the uninitiated—with names like 'Oligocene Bovey Beds' or 'Carboniferous Culm Measures'. The words in a geological map key may convey exciting information to people who have studied the subject, but beginners will sometimes find these maps quite beyond them.

Nevertheless, there is a lot of information to be picked up from a geological map by somebody who knows something about rocks, and knowing how to read one is important to anyone who wants to take the subject further. If you would like to learn how to read a geological map, and you don't wish to go to a geology class, there is an excellent, clearly-laid-out booklet for scholars called *Reading Geological Maps*, by M. J. Bradshaw and E. A. Jarman. It is published by the English Universities Press Ltd., and includes sections of coloured geological maps and information to help you to understand them in detail.

Essentially, there are no short cuts in the study of geology, but it is a gripping and fascinating subject for anyone who becomes involved in it.

As has been said, there are three main kinds of geological maps: 'solid', 'solid and drift'—and 'drift' maps which supplement the solid maps in some areas in the scale 1 inch to 1 mile series. The latter, when of coastline, sometimes include information about 'beach cobbles', as they put it.

The geological map of the British Isles, scale 25 miles to 1 inch, would be a useful map for anyone wanting some generalised information about the geology of the British Isles, mainly because the key is supplemented by easier-to-understand descriptions of the more esoteric geological terms.

There is a two-sheet version which does not include Ireland, scale 10 miles to 1 inch. Recent editions are over-printed to show sheet lines of the One Inch Series and Special District Maps. It follows that, because of the larger scale, a more detailed map can be given. This applies even more so, in the one inch series.

Because of damage by enemy action during the last war, many geological map plates were lost. The task of mapping the geology of Great Britain is under way at present, mainly on the 1 mile to 1 inch series. There are still gaps in the survey, but a large part of Great Britain has already been covered in this series, which are based on the Ordnance Survey maps.

These maps are obtainable from the Institute of Geological Sciences, from the specialist mapsellers, Edward Stanford, Ltd., 12 Longacre, London W.C.2, or from other geological map agents. Members of the public can visit the library at the Geological Museum and look at any of these maps, and geological maps of other countries, too.

In addition, the inexpensive handbooks of British Regional Geology, published by H.M.S.O., would probably be of considerable practical help to a beginner. They contain every sort of information about the rocks and mineral deposits in the area that the book covers, as well as a profusion of detailed geological maps, pictures of fossils in the district, and other relevant items. These handbooks are not expensive, and would add considerably to the interest of your hobby to have with you on a visit to an area, when you intend to go out pebble and rock hunting.

In rock hobby magazines, and catalogues from rock shops, there are almost always advertisements for small booklets about the area where the shop is, written by some local enthusiast, and printed by local printers. They generally cost from about 15 p. to 40 p., and, in more senses than one, they could be a 'mine' of information, for the rock collector. It might be worth sampling one or two about an area in which you are particularly interested.

A map *Gem Sites of Great Britain* is published and kept up to date by *Gems* (The British Lapidary Magazine), which is also publishing a book, giving rock sites in Britain. The magazine's address is included in the addresses at the end of the book.

PEBBLES AS YOU FIND THEM

When you first go out deliberately to collect pebbles, instead of happily and innocently picking them up because of their attractive appearance, as you formerly did, you are likely to stand helplessly looking at them in their apparently nondescript thousands, wondering which among them are the special ones.

The best thing to do is to go on as before, pick out which you think are the most interesting, but instead of throwing them into the sea, take them home to identify them.

The colours are brought out when the pebbles are wet, so you can tell something about a pebble by dipping it in the sea. You may well be aware that an agate usually has bands of whiter and darker colour in it, but it is equally likely to have a white skin—a porcelain-like coating acquired since it was formed, and in its travels. Even if you scratch hard at this skin you cannot always penetrate it to see the agate pure and clear inside. If the pebble tends to be pear shaped, or shows a knotty little hackly surface on any broken part, and you can see bands in it, there is a chance it may be an agate.

But it is an outside chance. You don't find many.

Large rock pebbles are often easy enough to identify by their visible characteristics. There are the speckles, the marks of the layers, the heavy, sugary-looking white. But what are they? You will easily be able to tell once you have learnt to recognise them, and you do this by practice. Sitting down with a few at home and assessing their features against charts like the ones in this book, or going to museums and looking at the specimens in the cases, with information about them written nearby.

You may find a round ball of stone, or another similar, curiously-shaped rock form, which, as it is plainly not a water-worn pebble, may puzzle you. Probably this is a concretion, rock which has formed round a nucleus, such as some shell or fossil, like the pearl which forms round a grain of sand in the oyster, or the round roe-like grains of oolitic limestone.

Concretions are hard, and sometimes hollow inside, with a beautiful lining of crystals. These are called 'geodes', and are usually cut in half for display. Geodes often occur in limestone. Cannon Ball concretions of dolomitic limestone are hard, heavy round balls which you could find lying about in an area where this rock occurs. The name is apt, as they do indeed remind you of cannon balls by their weight and shape. Some rounded concretions, such as marcasite nodules, reveal a radiating crystal structure with chrysanthemum-like metallic fibres when broken open.

One way of familiarising yourself with pebbles, and recognising a good specimen through a deceptively faded or discoloured coating is, if you own a tumbler polisher, to collect a batch of small- to medium-sized stones, not fussing too much about their singularity, and to polish about half of them—taking note of their original and polished appearance, and keeping the other half for comparison later. When you put them beside each other at the end of the polishing process, you will soon see which come out well, and acquire the knack of seeing the possibilities of an indifferent-looking weathered pebble, and also of judging, which really are rather dull, polished or unpolished.

By the end of a summer of collecting, examining and, perhaps, polishing your pebble finds, if you go back to the same beach where you began, you will congratulate yourself on how much more you know about pebbles.

SEDIMENTARY PEBBLES

These pebbles are derived from rock and formed from mud and sand; they range from soft to quite hard. You can scratch some chalk pebbles with a fingernail, but you would find it difficult to cut some limestone with a knife. Sedimentary pebbles tend to have a matt surface, and you can look for fossils and shells in them.

Chalk is the familiar powdery white rock from south-east England. It can be stained with impurities, and some kinds are cream-coloured. It has a dry feel, and wears away easily. The pebbles are usually little egg-shaped specimens, or you find rough broken lumps.

Limestone pebbles are smooth, clean-looking oval stones, flattish generally, and have the look of what is defined as stone. But

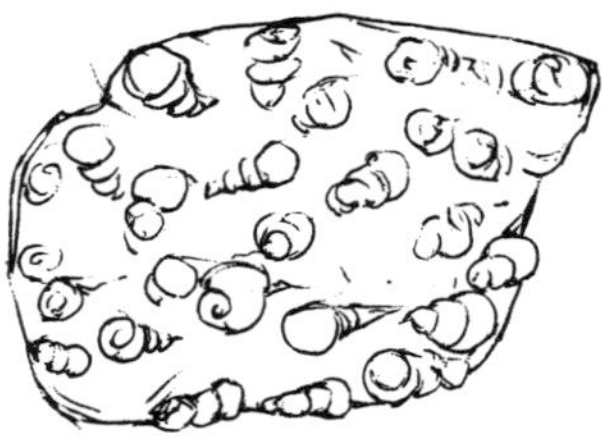

FIG. 2 Shelly limestone

some varieties of limestone are a mass of shells (shelly limestone, and blue-green Purbeck Marble) and of fossils (coral limestone, crinoid limestone). You may be able to see the fossils at a glance, or they may be too small to be seen with the naked eye and give the stone a speckly appearance. There is a beautiful example of fluted and ribbed coral limestone in the Natural History Museum in London.

Sandstone is not difficult to recognise. It looks like compacted sand, and is often a golden yellow or pale sand colour. Old Red Sandstone or Devonian, found in Devon and from Herefordshire and the Welsh Border is dark red, owing to its iron content. Look for a rain-pitted or rippled surface like the marks on sand under the sea. Other sandstones can be grey.

Mudstone looks like what it is: extremely fine textured clay that has become rock. It is mud-coloured. Shale, formed from compressed silt and clay, is dry-looking, and splits at its layered edges. It can be grey or black as well as brown.

Conglomerate pebbles are a mixture of lots of little, water-worn, rounded pebbles, which are bound together by a natural cement of rock called the matrix. You could almost mistake conglomerate for the pebble dash that builders use for rendering some house walls. Hertfordshire Puddingstone is conglomerate from north of London. It is possible for some conglomerates to contain huge pebbles and small boulders; others are made up of minute pebbles. Conglomerate can be distinguished from breccia, another mixture of stones in a matrix, by the fact that in breccia the stone fragments are always angular, and in conglomerate they are always water-worn and rounded.

Another sedimentary pebble that you might find is millstone grit, a rock which covers large areas of Britain, including the Pennines. Millstone grit is visibly sized grit particles in sandstone, and gets its name because it was used to make millstones.

A pebble with large crystals scattered in it is probably some form of porphyry, and is igneous, not sedimentary rock. Flint and chert, too, are not sedimentary rock pebbles; they are forms of the mineral silica which occur in sedimentary rock, namely chalk and limestone, and are normally found in an area of these rocks, or within water-borne distance from them, unless they have been carried for building materials elsewhere. If you look up at a chalk cliff or the steep sides of a chalk railway cutting, you can often see a black seam of flints, or several. Chalk is easily eroded by water and weather, and enduring flints outlast this softer matrix. You will find masses of flint pebbles on a shingle beach, and chert, practically indistinguishable from them, among them or near areas of limestone rock.

Nodules of flint, formed in cavities in the chalk, are knobby and extraordinarily branched, natural-stone sculptures; they are usually coated with a porcelain-hard white skin, through which you can see, where it is chipped, the opaque glossy black silica flint.

Flint was the sharp-edged stone used and shaped by stone-age man for his weapons and implements. It has a characteristic fracture, called 'conchoidal', which you can easily recognise by its shallow, scooped, concave-shell shape, sometimes small curved dents in the sharp translucent thin edge of the stone, sometimes larger fractures. Flint is brittle, and to knap, or break it is a skilled occupation.

It is usually grey, black or brown. Little, water-worn pieces of flint are usually irregular-shaped, hard stones, yellowish brown, or light brown, black, white-grey, with remnants of white coating, frequently with holes through them and fossil imprints. Rounded, oval flint pebbles are hard, smooth tough stones, shining and opaque where broken, and in a wide variety of shades of brown, white, grey, black, white-speckled, red, iron-stained or yellow. They polish well in a tumbler polisher, but it is difficult to get rid of all the white skin. Some flint is so translucent and pale that it could almost be mistaken for chalcedony. But this is not surprising, since it is made from the same mineral (silica) as chalcedony and quartz, and, like chalcedony, is crypto-crystalline, or smooth-textured.

Chert, found in limestone, is difficult to tell apart from flint, but it is usually lighter, and brown rather than black, or, often, a whitish-looking pebble.

IGNEOUS AND VOLCANIC PEBBLES

Granite has already been described, a speckled, granular, durable rock, either pink, grey or white, with black mica specks, and colourless quartz crystals, with the three minerals clearly visible. If the rock pebble looks like granite, but has relatively large quartz crystals in it, you have probably found some granite porphyry.

Porphyry is rock which has large crystals in a fine groundmass;

it is usually found in igneous rocks, such as porphyritic granite, or some volcanic rock porphyries.

A fine-textured, granite-like rock with mottled colours, green and white or black and white, could be diorite, which has a finely-speckled look.

Gabbro is found in the same regions as granite and looks very like it, but the mottling is large-textured and coarse, with no sparkling quartz crystals, ordinarily, and of dark grey, black and green minerals.

Quartz pebbles are found in association with igneous rocks. The commonest kind are round, humped and small discoloured pebbles with little red capillaries of iron impurities. Quartz can also be found in imposing large, smooth, well-shaped, snow-white pebbles. The semi-precious gem varieties of quartz are clear colourless (rock crystal); glassy pale pink (rose quartz); pale yellow (citrine); mauve to violet (amethyst) chevron-banded mauve (phantom amethyst); smoky quartz (smoked transparent); golden brown (cairngorm); black (morion). There are other varieties of quartz, which is one of the commonest of earth's minerals. You will be very fortunate to find gem pebble specimens in Britain, but they do occur. Quartz is extremely hard material, not to be cut with a steel knife.

Basalt, among the group of volcanic rocks, is black and lifeless looking, and to the eye, smooth textured. Obsidian, which is volcanic glass, is easily recognisable, because it looks like black bottle glass and has a concave, shell-like fracture (conchoidal) when broken.

Chalcedony, the name given to the collection of semi-precious gem minerals which occur in volcanic lava rocks, is itself tough blue-grey translucent material with a waxy, shadowy bloom, and sometimes milk-white bands. It can be found in grape-like clusters, or in sticks like stalactite formations.

The chalcedonies include agate, which has the same translucent tough texture and waxy bloom, but is often paler or more honey coloured, and distinguished by parallel bands of varying widths and spacing apart. Fortification agate has concentric angular bands; eye or orbicular agate has circular concentric bands; moss agate is chalcedony with what looks like moss inside, which is in fact a chemical inclusion; carnelian is flesh-red chalcedony; sard, similar to carnelian but darker red.

Other chalcedony agates include onyx, which is brown, black and white banded, with straight bands, and sardonyx, which has a similar banded appearance to onyx, but a red colour.

Water-worn agate pebbles do occur, and you can often find hackly fragments of chalcedony in areas where it is formed.

Jasper is a fairly common pebble, which is usually a smooth oval shape, slightly flat. It can be found in several strong, dark colours, the most well known being terra-cotta red or purplish-red; ochre and green are also colours of jasper. It is a rewarding pebble, its faded, weathered look as a water-worn pebble coming up strongly vivid and glossy when polished.

METAMORPHIC PEBBLES

In order to get some idea of the lovely colours and varieties of marble, which is metamorphic, recrystallised limestone, you would need to visit a museum collection, such as that of the Mineralogy department of the Natural History Museum in London.

Marble can be snow white. It can also be brilliant terra-cotta red (Duke's Red Marble, Derbyshire) or look like polished wood (Chatsworth Rosewood Marble, Derbyshire). Connemara marble is crystalline bright green; Tiree marble pink, with dark green flecks. In the rough of course, marble will look weathered and less colourful than polished versions.

Gneiss, as already said, is similar to granite, but instead of the black mica speckles being scattered evenly through the rock, as in granite, it is arranged in bands. The texture of gneiss is granular and hard, like granite, and it can be differentiated from another metamorphic rock, schist, by its coarser grain and pattern.

If you have found some slightly greasy-looking, lustrous, green, heavily-compressed and thinly-layered rock, you have probably got some mica-schist. Sometimes this rock is sandwiched with coarse-looking, slightly overblown-looking, irregular white crystals of pegmatite, a form of quartz-feldspar. Mica-schist, under certain conditions, produces garnet and other crystals.

Schist is characterised by fine banding and compressed layers, and can be pale grey, green, and sparkling. It is named according to its predominant mineral: mica-schist, hornblende-schist, etc. Another metamorphic rock, slate, is familiar to many of us, because it used to be widely used for roofing tiles. Slate can be grey, the dark elegant colour known as slate-grey. It is also sometimes red, green or black, and sometimes it is found in a combination of these colours, which merge together at a clear join of hue. Slate is fine-textured, and you can see the lines of the laminations down the sides, and flecks of shining mica which give it its sheen. Slate has an unusual feature, that it cleaves, that is, splits easily, at an angle to the direction of pressure. There is also a spotted form of slate. It makes beautiful large smooth, well-shaped pebbles, or thin sheets.

If you have ever visited the Lizard in Cornwall, you have almost certainly come across serpentine rock, which is green or red mottled, and is heavily worked for tourists. It makes smooth, almost polished, water-worn pebbles, not to be confused with very hard dark green, red-flecked bloodstone, or heliotrope, which is one of the chalcedonies.

Quartzite is recrystallised sand grains, which may be typically sand coloured, or white, or patched and mixed with colours. At first sight it looks like sandstone, but its hardness, compactness and solid smooth texture, and rather heavy feel, help to identify it. It is as hard as quartz.

OTHER KINDS OF PEBBLES

Strictly speaking, some of the specimens described here are not rock, but crystal. Brassy, gold-coloured shining iron pyrites, aptly called 'fool's gold' because it can look like a round knobbly lump of gold, or attractive gold nuggets of dice shaped crystals. Pyrites can appear as inclusions in other pebbles, as little flecks of gold.

Marcasite is a crystal found in chalk. It could look like a cannon-ball-shaped lump of stone when you find it, but if this is carefully split in two, you would find a silver sun-shaped formation, with needles radiating from the centre. Marcasite rusts quickly in air, so freshly-exposed crystals should be covered.

Galena is a glittering, heavy, bright silver crystal, in appearance rather like pyrites. Another kind of metallic mineral is hematite, which is kidney ore—a boiling mass of bulging lumps, and ferrous red, purplish black, or grey.

Amber, orange red-fossilised resin, and jet, petrified wood, which gives the name 'jet black' to intense, sooty black, are two other kinds of pebble you might find, normally on the east coast.

These descriptions are intended to help you assess what your pebbles are on the spot, when you find them. More detailed ways of identifying them with a variety of simple tests, are given with classified alphabetical charts later in the book. If you can't fit what you have found into any of these classifications, and you would like to identify it, the Institute of Geological Sciences (the Geological Museum) in London will give you free advice during certain hours. They have scientific equipment with which they can look at the crystal shapes in the rock, or the mineral content, etc. and expert knowledge. The Natural History Museum, London, also helps with advice on rock specimens.

CHAPTER THREE

Safety—and Equipment needed when Pebble-Hunting

Safety is a concern in rock hunting. And, although it may seem wildly improbable, there are also some possible dangers in the gentle art of pebble hunting. So, a few cautions are necessary for the beginners.

A FEW CAUTIONS

Hacking an embedded pebble out of a crumbly cliff face carelessly, might cause rocks to dislodge above you, which could fall and hit you on the head. You might, in an effort to reach some inaccessible cove, be cut off by a rising tide. This applies even more when exploring caves, by the sea or inland, where you could be dangerously vulnerable. Climbing down into flooded gravel pits or quarry workings has obvious hazards, too. If blasting operations are taking place, it is wise to respect warning notices.

As you go further afield and begin to collect in earnest, in remote places in the mountains or moors, you may venture into disused mine workings, which could be derelict and dangerous. If you have an accident, and are alone, the consequences could be serious, as they could be if you get lost underground searching in caves and potholes for mineral finds.

For this sort of expedition, you should always have at least one companion, and leave information where you are staying about where you are going (don't change your mind and go elsewhere!).

You should be properly equipped with good protective clothing, maps, torches and food.

Don't underestimate a trek on the wild moors five or ten miles long. You can be considerably isolated under these conditions, and changes of weather can be sudden. This is even more true when you go into mountains or are high up. This is the sort of expedition in which rules for mountaineers apply, and for this you would need a compass and other emergency equipment.

The other aspect of safety to watch is, when you are breaking rocks. Splinters can glance off when a rock is struck a sledgehammer blow—or even from a modest-looking pebble which you shattered thoughtlessly with another piece of rock. Quartz is harder than steel, and a fragment in your eye could not be extracted with a magnet as metal could be. If you or any bystanders watching you get a sliver of stone in the eye, the results could be very painful.

If you haven't got plastic goggles to wear when striking rock, cover it with a thick wad of newspaper or cloth before you hit it with your hammer. Hard hats are available for tough rock hunting operations, but in general it is useful to have a strong stitched hat to protect you from dust and small falling stones.

Ask Permission

It is always essential to ask permission if you wish to search for specimens on anyone's land. Falling out with people whose concern any territory is, is not pleasant, and not sensible either. Quarries would soon be closed to rock hobbyists, and regulations would be tightened if people collecting rocks and minerals became a nuisance in any way, and caused damage. At present, in this country at any rate, there is a tolerant attitude, and it is obviously wise to keep it that way.

EQUIPMENT—BASIC REQUIREMENTS

A small number of tools will be useful as you advance in your interest. For the beach all you need is a seaside bucket, a small penknife to test the hardness of the pebbles you are examining, and a magnifying glass. When it comes to breaking off or prising out rock specimens, it is a different matter. You will need some equipment for this, and certain tools, at least.

For the simplest requirements, the items in the first part of the list are all you need. For organised field trips into wild country a much wider list is included.

A penknife

It should have steel blades. You use it to scratch the weathered coating from a pebble to see the true colour underneath. It is also used to testing whether the stone will be hard enough to stand up to polishing, or simply to help you identify it on the spot.

A bag

You will require some container to carry your collection home in. A haversack is useful, because you can easily have it on your shoulder while you move about looking for specimens, for which it is helpful to keep both hands free. It should have a flap which fastens down securely, to prevent the pebbles from tumbling out when you least wish it, and a wide strap, possibly padded where it sits on the shoulders. A pocket on the outside holds small items stored separately; for example, your notebook and pencil and magnifying glass.

FIG. 3 Rucksacks

For ordinary purposes, a small bag is sufficient. If you are camping and want to pack clothes in it as well, you will want it to have two or three compartments and pockets, and to be very comfortable to carry.

It is wise to avoid having several different bags slung over you if you intend to walk over several miles of rough terrain. This can be very uncomfortable, and progressively more so as you travel. If you are taking a rucksack, get a good one, with a stable frame to support the weight, and deep outside pockets for storing specimens.

For ordinary purposes, if you own a small airline shoulder bag, made of canvas with several zipped compartments, this is practical for the pebble collecting purpose, though once again, make sure it is comfortable where it lies on the shoulder. Stones, as you can imagine, can weigh you down in no time.

If you want to add to your carrying capacity, a broad leather belt can be useful with canvas pouches fastened to it, in which you can store smaller items, leaving your pockets and haversack with slightly more space to spare.

Polythene bags and Tie-on labels

These are needed for keeping special specimens separate, and the haversack clean inside. The labels are to tie round the neck of the bag with the particulars of the place where the pebbles were found, and the date.

A notebook and pencil

Whatever you decide to make of the stones eventually, when you find them you should record details about where you picked them up, the date, and any other points of interest to record later in a card index of your collection, if you decide to make one.

TOOLS

Pebbles in your garden, or pressed into the ground by the side of the road are difficult to get out without the aid of suitable tools. You may need to poke away earth, or prise or lever it out, and will be glad to have a hammer and cold chisel to use. There are several sorts of hammer:

A crack hammer

This is needed for smashing rock. It is like a small sledge hammer, and you will need to own one only if you are a dedicated collector, or hope to make things for the home on your lapidary machine, out of slabbed rock.

A geologist's hammer or geologist's pick

This has a hammer head with a chisel or pick end. It is a handy tool to take out with you for collecting rock or mineral specimens.

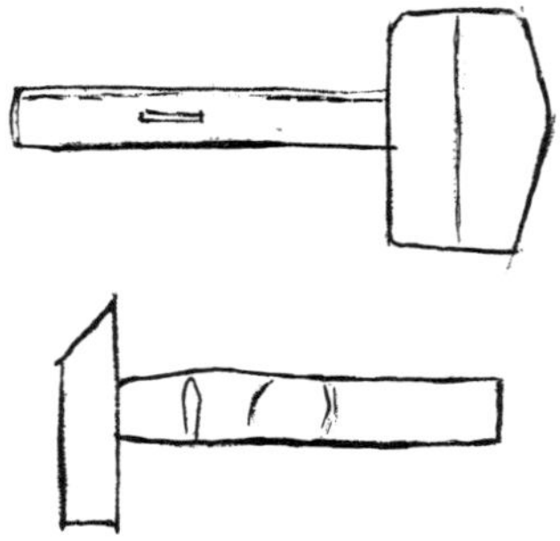

FIG. 4 Crack hammer and geologist's hammer

Cold chisels

These are bars of steel, pointed or flattened at one end. They are used for levering out embedded specimens, or to help break rock. You need to choose chisels with good leverage, the right thickness in proportion to their length for your purpose. Several sizes are useful: $\frac{1}{2}$ in.; $\frac{3}{4}$ in. and a small $\frac{1}{4}$ in. cold chisel for prising out small specimens.

Goggles

Inexpensive plastic goggles will protect your eyes when you fracture rock. Otherwise take newspaper to cover the rock when you hammer it.

Gloves

Hard-wearing gloves, gardening gloves or kitchen gloves will be useful to protect your hands from cuts and grazes when you are scrabbling among earth and broken rock.

Packing materials

Newspapers; cotton wool; tissue paper; and boxes for individual specimens which are especially fragile.

A magnifying glass

Magnifying glasses of different kinds are vital to the pebble and rock collector. A good-quality, three-inch diameter magnifying glass is needed for inspecting stones while polishing and cutting them during lapidary work, as well as when identifying them. Details of the equipment needed for lapidary work and jewellery making are given later, but you will need your magnifying glass at every stage.

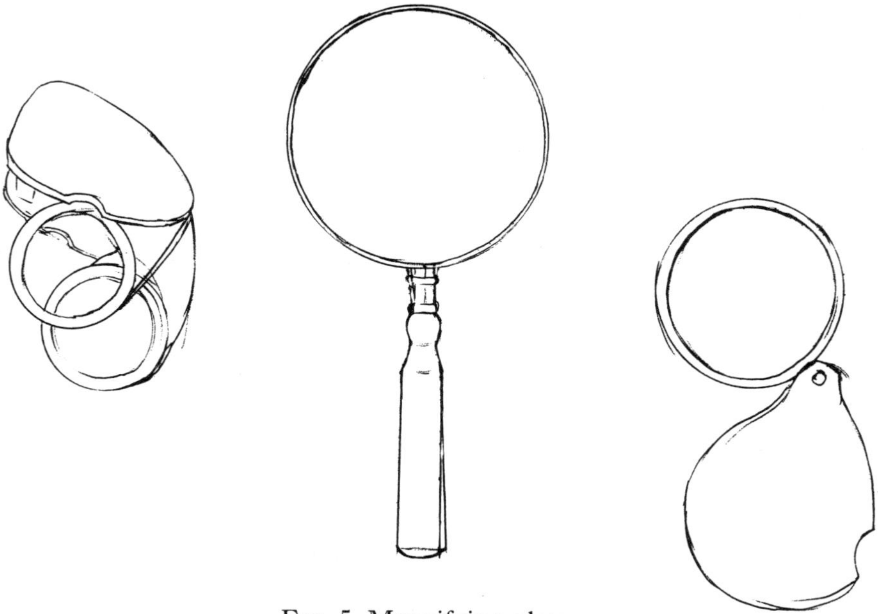

FIG. 5 Magnifying glasses

Pocket magnifying glasses, enlarging up to ten times, (or sixteen times, using the type with double lenses), have a small round lens which folds away inside a metal or plastic cover. They are held close to the eye and the pebble is brought up to the lens. Another type of pocket magnifier is a smaller-diameter version of the hand-held glass, which folds away in a leather cover.

FOR TRAVELLING FURTHER AFIELD

A few items have been mentioned already, such as a rucksack and crack hammer. For going into lonely areas off the beaten track, some other equipment will be essential.

Maps

Ordnance Survey maps in large-scale, local sheets of the areas you will be in, plus a small magnifying torch for reading the map in failing light.

A compass

In the small, over-populated British Isles, this may seem to be a ludicrous suggestion, when one thinks of the sign-posted motor roads and lanes everywhere. But mountains and wild areas of moorland are another matter, and this is where rock specimens are to be found, as much as on low-lying plains and lesser hills. To be lost high up in precipitous, lonely country in a mist is not funny. Any direction finder will be gratefully consulted under these circumstances.

Clothing

Hunting for pebbles, rocks and minerals is likely to keep you out of doors in all weather. You will need adequate weatherproof clothing, sun-proof as well as rain-proof, in the famous irregularities of the British climate, where it is common for the weather to change from one hour to the next. Having put on thin clothes for bright sunshine, you may well find yourself shivering in sleeting rain and a chill wind a couple of hours later.

Sensible tough denim or corduroy jeans and a jeans jacket will be comfortable in most weather, with the addition of woollen socks to pull over cotton socks, and a pullover for extra warmth when needed. A quilted waterproof anarak with a hood for outer clothing, and you will be wise to take a lightweight, strong waterproof cape or coat, too.

For a hot climate, to protect you from the sun, cool loose cotton clothing; a hat such as a stitched linen fishing hat which you can roll up and put in your pocket, and will also keep the dust of the rocks out or your hair and offer a small protection against flying stones.

Shoes

For finding pebbles on the beach, going in an out of the sea, you can dry out easily and be comfortable in canvas sandshoes worn over bare feet, in summer. For trekking into the hills, laced leather walking shoes, or even hiking boots, are necessary. Choose socks carefully for comfort, long enough and absorbent enough. If you will be going through scrub land, or nettles and brambles, it is wise to wear long trousers, to avoid being scratched and stung.

Really advanced rock hunters wear a hard hat, and approach

their job with caution. Hard hats can be bought from at least one supplier at present in Britain.

A waterproof torch

Sometimes you will be looking about in a cave or in damp gloomy conditions, you will want a rubber covered torch, with fresh batteries, if necessary. If not, at least take a pocket torch with you.

Water and Food

Five miles on foot over rough country, even in Britain, is different from five miles in a car. Carry drinking water in a plastic container, and a packet of water purifying tablets. Bread, cheese, hard-boiled eggs, unsquashable fruit such as apples, chocolate, raisins and nuts are sustaining foods.

First Aid Kit

Plasters are a minimum requirement. A small first aid kit designed for campers or motorists is useful to take with you.

N.R.C. Rescue Blanket

An item for your equipment that should be as basic as a first aid kit if you are going far off the beaten track, this is a lightweight foil sheet measuring 56 in. × 84 in. which will reflect back 90% of the body heat of the covered person, keeping out rain and snow. A type which you can only use so long as it is not torn, costs about £1.05; this is bright silver; another version of the blanket costing about £1.50 is orange on one side, silver on the other, and will reflect a searchlight, making finding the lost person easier. A reinforced foil blanket in silver, red and blue costs about £3.63. This can be re-used. Another protective item for emergencies is a large plastic bag, price about 60 p into which you can slip your sleeping bag and kit if you are lost and have to stay out all night. These blankets are supplied through mail order by Black and Edgington Ltd., Ruxley Corner, Sidcup, Kent, or from Blacks of Greenock Ltd., 22, Gray's Inn Road, London, W.C.1.

A lightweight tent

This will be much more comfortable if it has a lightweight groundsheet stitched to it. This can be done by sewing it on a domestic sewing machine, allowing a five-inch extra amount of groundsheet

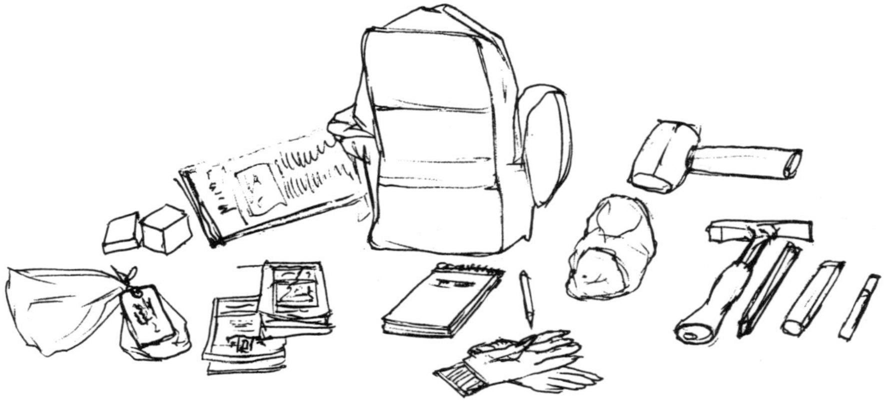

FIG. 6 Equipment used when collecting rocks and minerals

up the sides of the tent. This is only suitable for home-sewing on a cotton or nylon tent.

A box of matches and a tin opener

Always forgotten—but always needed!

Handbooks, geological maps

It is certainly advisable for anyone who is taking up the pebble hobby to gradually collect at least the British Regional Geology Handbooks, and to have one or more geological maps. The handbooks are available from Her Majesty's Stationery Office, and maps from the bookshop at the Geological Museum, Exhibition Road, South Kensington, London, or Edward Stanford Ltd. Mapsellers, 12 Long Acre, London, W.C.2, and other agents.

CHAPTER FOUR

Identifying your Pebbles

One benefit of the pebble hobby is that it exercises your powers of observation. You find yourself looking more attentively at nature in sharper focus. Once you have become interested in telling the difference between one rock and another, you begin to notice things about them, such as the way they shine or if they are scattered with crystals.

The vast majority of pebbles you pick up will be rock, with odds and ends of minerals. If you've had a lucky day, the collection that will tumble out of your haversack at home will probably consist of some pebbles and pieces of rock, a few fragments of chalcedony, perhaps one or two more prized specimens, a flint with a fossil in it, or a good piece of jasper. You may even have the pleasure of carefully unwrapping a small lump of glittering crystals from its cotton-wool-lined box.

The next thing to do is to try to identify the specimens. Identifying rocks is not quite so straightforward as identifying minerals, because many rocks are a mixture of several minerals. When a rock is composed of a single mineral, like chalk, then you can apply the mineral tests to it. Otherwise, to begin with, compare your rock pebbles with the rock identification charts.

There are simple ways of testing minerals at home using a few items you can probably find about the house without undue trouble. A copper coin, a penknife, a piece of broken window glass a couple of inches across and a steel file (from the toolbox, not the manicure set). You will also need a white bathroom tile, your magnifying glass, and pocket magnifier, too. And a notebook and pencil. It is best to work in a good light, at a table covered with newspaper, sitting comfortably.

First to give you a broad description of the sort of features of the pebble you will be looking at. Take one and hold it in the light, and look at it through the magnifying glass. Does it look glassy, or metallic? Can you see the shapes of crystals in it? Crystals are not only colourless; they can be tinted or dark green or black, microscopic or large.

Can you see your thumb through it? Can you scratch it with your fingernail? Scratching it with implements of different degrees of hardness will tell you where the specimen belongs in a scale of hardness. Are there any fossils in it? Bands of colour? Does it feel relatively heavy for its size?

When you work your way through the methods of identification jot the results down as you go along, so you can match them with the charts more quickly.

IDENTIFICATION METHODS

The features looked at when identifying minerals are: colour; hardness; transparency; lustre; fracture (the way they break); streak (the colour of the mark the mineral makes when drawn across the back of a bathroom tile); cleavage (the way the mineral cleaves, or splits, naturally); structure (whether the mineral is built of crystals, called 'crystalline', or whether it has a non- or micro-crystalline texture, which looks as smooth as syrup, and is called 'crypto-crystalline'.

These tests can be carried out quite easily by anyone at home with a few simple items mentioned earlier. They can, of course, be carried out more exactly in a laboratory using precision scientific equipment, and mineralogists would also test other things such as specific gravity, the weight of the mineral compared with the weight of water; refraction of light; the crystal structure.

But for average purposes you can go a long way with the homely tests. Try working your way through the following tests with two or three mineral samples, such as flint, or chalk.

Colour : a useful guide, but because minerals often come in a wide range of colours, only relatively useful. Nevertheless, do make a note of the colour.

Hardness : each mineral has a hardness number which corresponds to a grade of hardness in Mohs' Scale. This is a scale of hardness from 1 : talc (the softest), to 10 : diamond (far and away the hardest). How do some everyday items fit into this scale of hardness ? On the

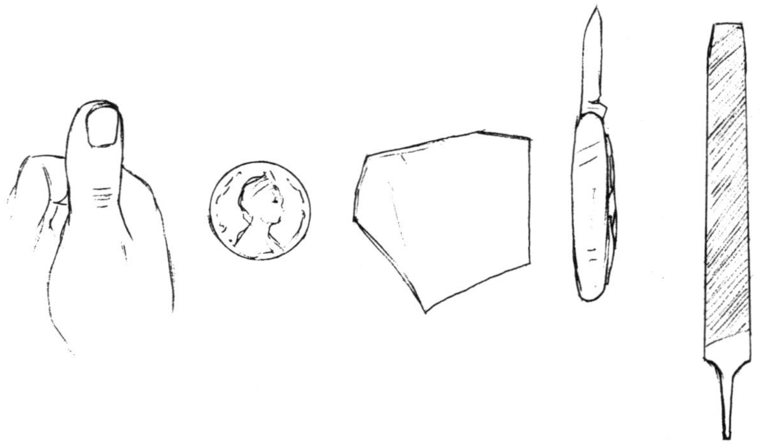

FIG. 7 Items used for testing hardness

scale, a fingernail = $2\frac{1}{2}$; a copper coin = 3; window glass = $5\frac{1}{2}$; a knife blade = 6; and a steel file = $6\frac{1}{2}$–7. If you can scratch the mineral with your fingernail that is, make a visible incision in it, then you know it is less than hardness $2\frac{1}{2}$, and can look among minerals in this category in the identification charts. If you can scratch it with a copper coin, but not your fingernail, you will know it has about hardness 3. If you can scratch it with window glass, but neither with your finger nail, nor a copper coin, you will know it comes between hardness 3 and $5\frac{1}{2}$, and so on.

Mohs' Scale (named after Friedrich Mohs, a German mineralogist who invented it in the last century).

1	Talc	
2	Gypsum	Fingernail $2\frac{1}{2}$
3	Calcite	Copper coin 3
4	Fluorspar	
5	Apatite	Window glass $5\frac{1}{2}$
6	Feldspar	Knife blade 6
7	Quartz	Steel file $6\frac{1}{2}$–7
8	Topaz	
9	Corandum	
10	Diamond	

(Write down the hardness number for your specimen)

Transparency: the way in which you can see light through the mineral in the following three degrees:

Transparent: can you see your thumb through it when you hold it up to the light?

Translucent: can you see dim light through all the stone, but not objects?

Opaque: you are unlikely to see anything through an opaque stone, except possibly a glow through the thin edges.

(Write down the transparency beside the hardness number for your specimen.)

Lustre: the way the mineral reflects light. This applies to minerals and rocks of a single mineral, not mixed mineral rocks.

Vitreous: glassy.

Waxy: a soft waxy glimmer.

Silky: with the sheen of silk or satin.

Resinous: like hardened sap or heavy gum.

Pearly: opaque with a milky shimmer.

Metallic: has a metallic lustre.

Adamantine: hard, brilliant reflection of light, like diamond.

(Write the lustre of your specimen beside the hardness number, and the transparency.)

Other effects of light on the specimen:

Fluorescence: a glow given off when the mineral is affected by ultra-violet rays.

Phosphorescence: the mineral continues to glow after the rays have been removed.

Irridescence: inner shimmering in the mineral reflecting colours. Opal is irridescent.

Chatoyance: the mineral catches the light like a cat's eyes at night. Tigereye is chatoyant.

Asterism: the light reflects in the mineral in star-like rays.

(N.B. You are more likely to find specimens reflecting light like this overseas, but minerals do exist with this feature in Britain. If you have found a specimen, note this too.)

(Note down any of the above effects when writing the record of your specimen.)

Streak: the colour the mineral makes when drawn across an unglazed reverse side of a white tile.

(Only a few minerals have a noticeable streak. Note it beside the other characteristics if it is marked.)

Fracture: the characteristic way the mineral breaks.

Conchoidal: a shallow, scallop-shell-shaped break, scooped out with ridges across it.

Hackly: a gnarled, knotty fracture.

Uneven: a rough fracture.

Even: a smooth fracture.

Splintery: a fibrous, splintery fracture.

(Note the fracture type beside other results.)

Cleavage: the way the mineral breaks naturally, a tendency to part in two easily along smooth planes, when tapped correctly. Experts can tell from the angle of cleavage which the mineral is, taking other features into account, but this is rather a complicated process for amateurs who have not got a fair amount of knowledge about crystal structures.

Specific gravity: deciding the specific gravity of a mineral, that is, its weight compared to water, is a scientific exercise, requiring some scientific equipment, including an accurate scale balance. The specific gravity values have been included in these charts for anyone who knows how to carry out these scientific tests, but also because they convey something about the rocks and minerals without scientific tests.

The smaller figures in the scale, i.e. jet's specific gravity of about 1·10–1·40, indicate a substance which is light in proportion to its size. The heaviest substance in these tables, in proportion to size is galena, ore of lead, and this has a specific gravity of about 7·4–7·6. A glance at the specific gravity figure will soon help you to decide whether the specimen you are identifying is light or heavy for its size, and fits roughly with what you would guess from its specific gravity value, in relation to other specimens.

Other tests: there are some other ways of identifying minerals at home, such as testing some minerals with dilute hydrochloric acid to see if they effervesce. There are also other varieties of lustre, fracture, etc. But those given here are the main ones for ordinary purposes.

CHART 1—ROCK PEBBLES AND ROCKS

Some rocks consist largely of a single mineral; others are a mixture of minerals.

name, type	*description, colours*	*hardness, fracture, sp.gr. lustre, etc.*	*other characteristics*
Alabaster sedimentary	Looks like a lump of camphor or cooking salt, or snow. Is pure gypsum rock. Can be streaked brown or green.	H.2; sp.gr. 2·3; opaque.	Can be polished by hand with care, but too soft to tumble polish.

name, type	*description, colours*	*hardness, fracture; sp.gr.; lustre, etc.*	*other characteristics*
Basalt igneous, volcanic	Black or very dark green, lifeless looking rock. Can be scattered with green olivine crystals.	Various	Weathers brown. Is smooth-textured; will polish.
Breccia sedimentary	*Angular* chips of various stones in a matrix (natural cement) of limestone or other rock. Various colours.	Various	If matrix is hard, will polish.
Chalk sedimentary	White, powdery rock, grades in some zones into other rocks with yellowish grey colours. Chalk is typically, white, when freshly broken.	H.3; sp.gr. 2·72; opaque; white streak.	Flints and marcasites are found in chalk, and shells and fossils. Fizz in dilute acid. Not suitable for polishing.
Conglomerate sedimentary	*Rounded or ovoid* pebbles mixed and cemented in a matrix. Variety of colours.	Various.	If with hard matrix, will polish.
Diorite igneous	Looks rather like fine-textured, mottled granite. Speckled black or dark green and white, grey-black or grey-green.	Various.	—

name, type	*description, colours*	*hardness, fracture; sp.gr.; lustre, etc.*	*other characteristics*
Dolerite (also called micro-gabbro) igneous	Slightly 'sandy' textured, black or dark-coloured rock. Can also be dark or grey-green, or speckled green and grey-white quartz-dolerite.	Various.	Any crystal inclusions shine, otherwise dull.
Dolomite (magnesium limestone) sedimentary	White, yellow, grey, green, brown, etc.	H.3·5; sp.gr. 2·8–2·9; opaque	Concretionary: spherical formation. Sometimes lined with pearly crystals. Also in crystalline form.
Feldspar The commonest mineral; rock-forming in many rocks.	Pink or white (orthoclase) or grey or grey-white (plagioclase)	H.6·65; sp.gr. 2·5–2·75; transparent to opaque; vitreous to pearly like china. Cleavage: two ways.	Also gem feldspar; moonstone, sunstone, etc.
Gabbro igneous	Looks like very broad patterned granite, but without the shiny quartz crystals. A mixture of black/grey/green.	Various.	—
Gneiss (pronounced 'nice') metamorphic	Mineral ingredients often similar to granite, but altered so they have a speckly black stripe, 'zebra' banded appearance.	Various.	Folded gneiss shows folds in banding.

name, type	*description, colours*	*hardness, fracture; sp.gr.; lustre, etc.*	*other characteristics*
Granite igneous	Hard, granular rock with densely-packed mixture of three visible minerals: pink, white or grey feldspar, colourless glittering quartz crystals, and shiny black biotite mica flecks.	Various.	Can be slabbed and polished.
Hornblende rock-forming mineral	Usually found as ingredient in rocks. Dark-green, black.	H.5–6; sp.gr.: 3·3.	—
Laurvikite igneous	Laurvikite, from Norway, is a decorative rock; polishes with a rich blue shimmer, a play of light called Schillerisation	H.6.	Pebbles found on east coast in boulder clay.
Limestone sedimentary	Attractive stone, with a matt texture, in a wide range of stone colours: white, cream, yellow, blue-green, gun-metal grey, almost black. Often thick with shells or fossils.	Varies from soft to fairly hard. Streak: white.	Weathers easily. Fizzes in dilute acid. Made of same mineral as chalk. Chert forms in limestone.
Forms of limestone:			
Coral Limestone	Fossils of coral in limestone, ribbed fluted, scattered.	Various.	Similar.
Crinoid Limestone	Contains fossils of sea lilies, called crinoids.	Various.	Similar.

name, type	*description, colours*	*hardness, fracture; sp.gr.; lustre, etc.*	*other characteristics*
Oolitic Limestone	Limestone which appears to consist of the hard roe of a fish. Tiny round concretions, sometimes easily visible, sometimes too small to see without magnification.	Various.	Similar.
Purbeck Marble	Blue-green limestone from the Isle of Purbeck in Dorset. A mass of shells, sometimes too small to see without a magnifier.	Various.	Similar. (A similar limestone is Sussex Wealden Marble).
Pea Grit Limestone (pisolitic limestone)	Similar to 'oolitic' limestone, but much larger rounded concretions.	Various.	Similar.
Shelly Limestone	Limestone packed with shells of snails and other shelly sea life.	Various.	Similar.
Malachite rock-forming mineral	Bright green with massive swirls.	H.4; sp.gr. 3·7–4; opaque; uneven fracture; brittle.	Botryoidal form, like clusters of bubbles, or massive.
Marble metamorphic limestone	Recrystallised limestone, white or marbled in various colours. Tiree Marble: pink with dark green specks.	Various.	Soft but polishes and carves well (though not in tumbler polisher).

name, type	*description, colours*	*hardness, fracture; sp.gr.; lustre, etc.*	*other characteristics*
Marl sedimentary	Brittle, dry looking, opaque clay rock, with rough layers; chips off easily.	Various	Shelly marl found in marshy areas. Keuper Marl, white with brown, iron-oxide-stained bands. Also red.
Mica rock-forming mineral	Transparent and peels off in thin leaves. Used for stove windows. Colourless (Muscovite); black (biotite).	H.2–2·5; sp.gr. 2·8–3·1. Transparent to opaque. Lustre vitreous-pearly. Sometimes silky.	—
Millstone Grit sedimentary	Sandstone with visible particles of grit; used for making millstones.	Various	—
Mudstone sedimentary	Very fine-textured clay-rock in colours typical of clay: grey, whitish, brown.	Various	—
Obsidian igneous (volcanic)	Black volcanic glass. No visible crystals.	Conchoidal fracture, sharp edges; vitreous lustre. H.5; sp.gr. 2·3–2·5.	Pitchstone, U.S.A. red or brown volcanic glass. Snowflake Obsidian seems to have snowflakes in it. Will polish.
Olivine rock-forming mineral	Olive and other shades of green.	Vitreous lustre: H.6·75; sp.gr. 3·27–3·37.	Gem Olivine is called peridot.

name, type	*description, colours*	*hardness, fracture; sp.gr.; lustre, etc.*	*other characteristics*
Porphyry	A form in rocks which consists of large crystals scattered in the groundmass. Crystals vary in size and shape.	Various	i.e. granite-porphyry; basalt-porphyry.
Quartzite metamorphic sandstone	Quartzite is recrystallised sand particles; it can look like dense, hard sandstone, be patchy, or appear to have runs of yellow and red paint in it. (Type: from West Country).	Very hard, as hard as quartz; H.7.	Heavy stone.
Sandstone sedimentary	Sand grains compacted by clay or limestone. You can scratch particles of sand from some sandstone with your fingernail, some is tougher. Sand-coloured, or grey.	Various.	Can show rain pits and ripple marks on surface. Old Red Sandstone, dark red, even purplish-red, owing to iron content. From Devon, Hereford and Welsh Border.
Schist metamorphic	Rock compressed in layers. Sometimes silvery-green mica-schist, pale green, fine-grained rock with darker bands. Peels and cleaves in layers.	Various.	Called by its main mineral ingredient: mica-schist; hornblende-schist.

name, type	*description, colours*	*hardness, fracture; sp.gr.; lustre, etc.*	*other characteristics*
Serpentine metamorphic	Named because of its serpent look, green mottled with red; or red mottled with dark green.	H.3–4; sp.gr. 2·5–2·65; Fractures unevenly.	Fibrous. Cuts easily. Rather soft for tumble polishing.
Shale sedimentary	Very fine silty textured rock, in whitish colour, mud colours. Looks like slaty-clay, cracked and split, roughly shelved and split down sides.	Various.	Black when it contains oil.
Siltstone sedimentary	Looks like matt hard mudstone or clay	Various.	
Slate metamorphic	Shale metamorphosed, can be slate grey, green or red, or black. Flakes of mica give slate its sheen. Some slate spotted in self-colour. Slate is fine-textured, and shows lamination lines down sides.	Various.	Split and used for roof tiles; cleaves at right angles to pressure direction.

FIG. 8 Malachite

FIG. 9 A geode

Fig. 10 A marcasite nodule

CHART 2—MINERAL PEBBLES AND MINERALS

name, type	*description, colours*	*hardness, fracture; sp.gr.; lustre, etc.*	*other characteristics*
Amber fossilised resin	Pale yellow, or stronger yellow with lurid reddish-brown orange. Sometimes contains insects.	H.2–2·5; sp.gr. 1·05–1·10; resinous lustre, with a glossy, lumpy appearance. Translucent.	Amorphous, that is, with no definite internal structure. Carves and polishes well by hand. (Do not tumble.)
Blue John fluorspar	White glassy crystals with dark, purple-blue bands.	Translucent; H.4; sp.gr. 3–3·25; brittle, fragile, heat-sensitive.	Becoming rare. Mined Castleton, Derbyshire. Cubic crystals. Has to be cut and polished with extreme care.
Chalcedony a common form of silica	Tough, bluish or grey mineral sometimes with milk white bands. Also green imported variety.	Translucent; waxy; H.6·5; sp.gr. 2·6. Fibrous splintery fracture.	Crypto-crystalline, (smooth-textured). Found in bubble-like solid clusters or stick stalactite. Forms in lava cavities.

name, type	*description, colours*	*hardness, fracture; sp.gr.; lustre, etc.*	*other characteristics*
Forms of chalcedony:			
Agate	Agate can be *parallel banded*; *eye* or *orbicular* (agate with bands in concentric circles); *fortification* agate with angular bands; *moss agate*, with what appears to be moss inside but which is in fact green earth inclusions; *heliotrope* or *bloodstone*, dark-green chalcedony with red, blotchy spots.	As chalcedony. Hackly fracture.	Porous, easily stained. Can be tumble polished.
Carnelian, or cornelian	Flame-red chalcedony; *sard* is similar to carnelian, but a darker colour.	Similar.	Carved as seals in antiquity.
Onyx	Agate with straight bands; brown, black and white. Often dyed.	Similar.	Can be cut as cameos.
Sardonyx	Like onyx, but reddish-brown, and with white, parallel straight bands.		
Feldspar crystalline type. (see rocks)	White, pink, yellow, red, green, grey, brown, blue-green	H.6; sp.gr. 2·56–2·75; translucent to transparent. Vitreous-to-pearly lustre.	Can be polished.

name, type	*description, colours*	*hardness, fracture ; sp.gr. ; lustre, etc.*	*other characteristics*
Flint and Chert silica	*Flint* is found in silvery grey, black, brown; often in form of nodules with a porcelain like white skin. Sometimes red, iron-stained or yellow.	H.7; Conchoidal fracture; opaque with translucent edges—the sharp edges that were stone-age tools and weapons.	Crypto-crystalline (smooth textured). Forms in chalk. Sometimes nodules have crystal lined cavities.
	Chert is a lighter, browner, white or greyish coloured silica, so like flint as to be difficult to tell them apart.		Chert forms in limestone, sometimes in concretionary masses. Both will polish.
Fluorspar fluorite	The crystal form to which Blue John belongs. White, banded with yellow, brown, red, or blue-green.	H.4; sp.gr. 3·17–3·19; brittle, heat-sensitive; handle and polish with care.	Cubic crystals.
Hematite (kidney ore)	Ferrous red, grey or black mineral, which is formed in a mass of bulging, round lumps (called 'reniform').	H.6; sp.gr. 5·2; dull metallic lustre; fibrous fracture; streak: red.	Do not include even a small piece in a tumbler load, or you will find your stones grinding in thick red soup. Very messy to polish.
Jasper	Red, brown, yellow or green, impure silica. Also banded varieties.	H.7; sp.gr. 2·66; opaque; dull lustre. Even slightly-earthy texture.	Fairly common as well-shaped, water-worn pebbles. Polishes well.

name, type	*description, colours*	*hardness, fracture; sp.gr.; lustre, etc.*	*other characteristics*
Jet fossilised wood	Intensely black lightweight fossilised wood found in seams at Whitby, Yorkshire, and sometimes on beaches of east coast.	H.3·5; sp.gr. 1·10–1·40.	Carves and polishes well. Victorian favourite.
Quartz common form of silica	See below.	H.7; sp.gr. 2·65–2·66; transparent to opaque; fracture: uneven, conchoidal. Lustre; vitreous.	Crystalline. Found as crystals or water-worn pebbles.
Some varieties			
Amethyst	All shades of violet; also chevron-banded amethyst: 'phantom amethyst'.	Similar.	Similar.
Cairngorm	Golden-brown, clear quartz.	Similar.	Similar.
Citrine	Clear yellow quartz.	Similar.	Similar.
Milky quartz	Milk-white, opaque quartz.	Similar.	Similar.
Morion	Very dark-brown, clear quartz.	Similar.	Similar.
Rock crystal	Colourless, transparent.	Similar.	Similar.
Rose quartz	Delicate, transparent pink quartz.	Similar.	Similar.
Smoky quartz	Light-brown, clear quartz.	Similar.	Similar.

name, type	*description, colours*	*hardness, fracture; sp.gr.; lustre, etc.*	*other characteristics*
	Small creamy pebbles with red capillaries due to iron impurities common. Other varieties: Green aventurine, rutilated, catseye and tigereye; mostly imported in Britain.		
Satin-spar gypsum	Satin-silk white shimmering spar.	H.1–2; sp.gr. 2·2–2·4; Silky lustre.	Very soft, so grind carefully, and hand polish.

CRYSTALS

While you are collecting pebbles and looking at rocks, you will, sooner or later, almost certainly come across some crystals—besides those which form part of the rocks. They may be the sparkling lining of the hollow in a stone, or a glittering vein running through it. You could find them in a cluster of geometrical shapes, or as individual crystals scattered in a rock matrix, such as garnets. Diamonds, sapphires and emeralds are found as crystals in a rock matrix.

Crystals are blocks of mineral which have flat surfaces, called faces. They grow from microscopic to enormous sizes, and in six main crystal systems develop innumerable permutations on their basic definite forms. One crystal system is cubic, the form of a die; another the matchbox shape; another, with six equal sides, the hexagonal form of quartz crystals (which also have a six-sided pyramid at one or both ends). They form endless elaborations on their basic form.

Crystallography is a complex, fascinating subject, but models of crystal structures, and good examples of the crystals themselves, can be seen in the Natural History Museum, London, in the

Mineralogy Department, and they are presented in an orderly, simple way which makes it easier for an interested amateur to understand. Most popular books about geology contain a section on crystals, and you can, of course, read specialist books or the appropriate section in the encyclopedias in your local reference library.

MINERAL GEMS AND CRYSTALS

It is possible to find a surprising variety of crystal gems in the British Isles, though not often of a size which can be faceted and polished for jewellery. To qualify as gems they need to be at least hardness 6. Nevertheless, a piece of showy calcite crystals, mounted or not, makes a fine mineral sculpture, alongside rarer species.

CHART 3—MINERAL GEMS AND CRYSTALS

name, type	*description, colours*	*hardness, fracture; sp.gr.; lustre, etc.*	*other characteristics*
Barytes heavy spar	A noticeably heavy mineral, in bulky massive form, or as crystals. White, tinged with yellow-green, red or brown.	H.3; sp.gr. 4·5; lustre: vitreous to resinous, in the crystalline form; earthy, dull, in the massive form	—
Beryl	Usually pale green. Gem form, emerald, bright green, aquamarine, light blue, and other tints.	H.7·5–8; sp.gr. 2·65–2·85.	Hexagonal crystals. Elsewhere, sometimes found as pebbles.
Blue John	White, glassy crystals with dark, purple-blue bands.	H.4; sp.gr. 3–3·25; Translucent; brittle, fragile, heat-sensitive.	Cubic crystals; becoming rare. Cut and polish with extreme care.

ROCK PEBBLES

1 Mica schist 2 Quartzite pebbles 3 Quartz pebble 4 Schist 5 Granite (rock and a pebble) 6 Limestone 7 Quartz veined sandstone 8 Cannon Ball concretion (dolomitic limestone) 9 Mica schist 10 Polished limestone 11 Slate 12 Sandstone

ROCK PEBBLES

1 Hertfordshire Puddingstone (conglomerate) 2 Hematite 3 Shale 4 Flint and chert pebbles 5 Breccia pebbles 6 a: silvery galena; b: brassy iron pyrites 7 Red and green serpentine pebbles 8 Chalk pebbles 9 Polished and unpolished basalt with olivine crystals 10 White gypsum marl and dark shale

VARIOUS CHALCEDONIES, AGATES ETC. (polished and unpolished)

1 Tasselled amber pendant, and beads 2 Agate beach pebbles 3 Carlnelians 4 Chalcedonies 5 Agates in jewellery (including fortification agate and onyx) 6 Bloodstone (heliotrope) 7 Jet brooch 8 Jasper 9 Banded jasper 10 Moss agates

A FLINT

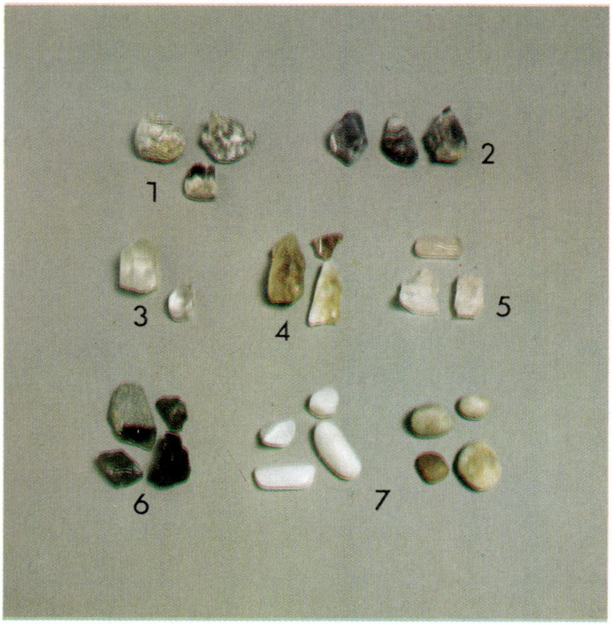

QUARTZ (polished and unpolished)

1 Phantom amethyst beach pebbles 2 Mauve amethyst 3 Rock crystal 4 Citrine 5 Rose quartz 6 Smoky quartz 7 Two groups quartz pebbles

name, type	*description, colours*	*hardness, fracture; sp.gr.; lustre, etc.*	*other characteristics*
Calcite	A common rock-forming mineral (chalk and limestone are calcite) which is usually white, but can also have yellow, green, violet, red, brown, blue or black tints.	H.3; sp.gr. 2·72.	Crystal forms include nail-head spar, like pile of nail-heads, and dog-tooth spar, pointed crystals.
Dolomite pearl spar	White, sometimes yellow-tinged, pearly lustre crystals.	H.3·5–4; sp.gr. 2·8–2·9.	—
Feldspar	White, pink, yellow, red, green, grey, brown, blue-green.	H.6; sp.gr. 2·56–2·75; translucent to transparent, vitreous to pearly lustre.	Can be polished.

FIG. 11 Monoclinic crystal —e.g. feldspar

FIG. 12 Quartz crystals

FIG. 13 Crystals in a matrix

FIG. 14 Fluorite crystals

name, type	*description, colours*	*hardness, fracture; sp.gr.; lustre, etc.*	*other characteristics*
Galena an ore of lead	Bright silver to lead grey.	H.3·5; sp.gr. 7.4–7.6; Metallic lustre.	Cubic crystals, or massive granular. Notice softness and heaviness.
Garnet	Crystals in rock matrix. Various types, mainly in wine-red colours, but also lime green.	H.6·5; sp.gr. 3·4–4·3.	—
Marcasite	Form in rough spherical nodules in chalk, which when halved, reveal the silvery radial crystals of marcasite.	H.6–6·5; sp.gr. 4·8; Metallic lustre.	Rusts easily and rapidly when exposed to air.
Pyrites (fool's gold)	Brassy, gold-coloured cubic crystals or nodules. Flecks and veins of pyrites in other minerals.	H.6·5; sp.gr. 4·84–5·10; Metallic lustre.	—
Selenite gypsum	Colourless transparent gypsum in crystal form.	H.1–2; sp.gr. 2·1–2·4.	—
Topaz	Yellow or colourless gem crystals in rock matrix, also pale blue.	H.8; sp.gr. 3·5–3·6.	Topaz Scotch Pebbles are water-worn lumps of the gem found in Scottish rivers.

name, type	*description, colours*	*hardness, fracture; sp.gr.; lustre, etc.*	*other characteristics*
Tourmaline	Stick-like, flat-sided crystals scattered and embedded in a rock matrix. Glittering black, or elegant rainbow tints, including pink into green.	H.7; sp.gr. 3–3·2.	Schorl: black tourmaline crystals, common in Cornwall.
Turquoise	Brilliant light blue to intense bright blue, and blue-green.	H.6; sp.gr. 2·6–2·85; Earthy blue in matrix, or waxy stone; opaque.	Amorphous, smooth texture.
Zircon	Red, yellow or green crystals, or water-worn fragments.	H.7·5 sp.gr. 3·95–4·72; adamantine lustre.	—

SHELLS AND FOSSILS

Shells and fossils are as much a part of the pebble hobby as crystals, since they form the material which is a substantial ingredient of various sedimentary rocks. Under the microscope chalk reveals masses of minute shells and fossils. Shelly limestone takes its name from the abundance of shells and fragments of shells that give it its attractive appearance. Crinoid and coral limestone are distinguished by their fossils, which look so decorative when polished.

Fossils are the remains of living creatures which existed at various stages during the millions of years of the Earth's history—hard bones and teeth, and the carapaces of long-gone creatures that lay in mud and sand so long that they became part of the rock itself. Sometimes their forms are in relief, and often they have left casts; petrified fossils are those in which the original material has

FIG. 15 A fossil shell

FIG. 16 Fossil Ammonites

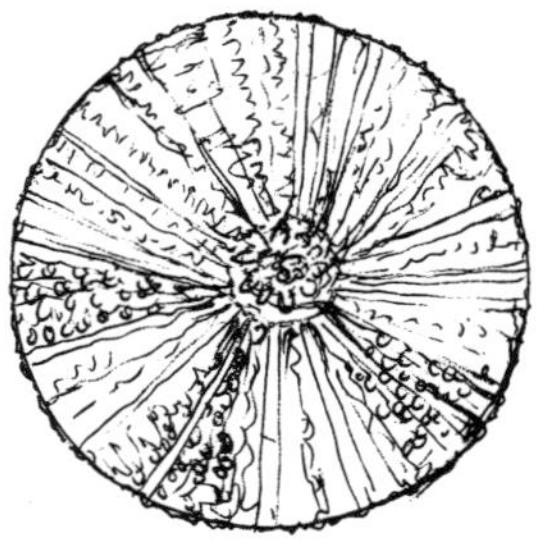

FIG. 17 Echinoid fossil

been replaced by iron pyrites or silica, acting as a natural filling in the mould. Leaves slowly rotting have left imprints, and worms their trails; prehistoric reptiles have left their skeletons and even footprints.

Fossils can only survive in sedimentary rocks. The molten heat of igneous rocks would destroy all remains, as would the vast pressures of metamorphic process, which at most would leave crushed fragments.

To geologists, fossils are the record of life on earth. In each geological era, fossils were laid down, layer by layer, stage by stage. Therefore, a fossil destroyed by unscientific handling, can be a loss of information to our accumulation of knowledge. This may not have a lot of significance in respect of common fossils—sponge spicules and graphtolites and fragments, but if some unique fossil link in the story of evolution was removed from its site without a proper record being made of where and which way up it was when it was found, with possible photographs and sketches to supplement the description, it could mean that irreplaceable geological evidence has been damaged or destroyed.

The Natural History Museum in London issues a pamphlet, Instructions for Collectors, No. 11; Fossils, Minerals and Rocks, price 10p. which tells a prospective fossil collector what to do, and what not to do.

Shells of land and sea creatures when fresh can be recognised by their coloured, bony appearance.

IMPORTED STONES

The rocks and minerals found in Britain are the same as you would find of the species anywhere, with the exception of Blue John fluorspar, which is only mined in England. Quartz is quartz where-ever you find it; so is agate.

There are stones you would not be able to find in Britain, and the rock shops import quantities of the semi-precious variety. From India, Brazil, Australia, Africa, the mined chunks are brought in: jadeite, amazonite, tigereye. They are sold by the piece or the pound, and although they are more expensive than finding your own, to buy them already worked by a jeweller would cost vastly more. Gems that our grandmothers used to prize and display are there for you to make up in your own workshop. Indeed, you are far more likely to find suitable specimens of fine quartz to make up or tumble polish this way, than any other.

CHART 4—IMPORTED STONES

name, type	*description, colours*	*hardness, fracture; sp.gr.; lustre, etc.*	*other characteristics*
Azurite	Intense, strong blue counterpart of malachite.	H.3·5; sp.gr. 3·77–3·89.	—

name, type	*description, colours*	*hardness, fracture; sp.gr.; lustre, etc.*	*other characteristics*
Chrysoprase	Green chalcedony.	H.6·5; sp.gr. 2·6.	
Feldspar varieties:			
Amazonite	speckled blue-green, a pale clear colour.	Similar, but lustre varies.	—
Labradorite	Blue-grey with shimmer of colour	Similar, but lustre varies.	—
Moonstone	yellow, milky-colourless.	H.6; sp.gr. 2·56–2·75; opalescent lustre.	—
Sunstone or Aventurine	Sparkling, spangled green or bronze.	Similar, but lustre varies.	—
Jadeite	The jade carved by the Chinese was jadeite or nephrite. Jadeite is granular, white, various greens, red, lavender, mauve, orange.	H.6·5–7; sp.gr. 1·65–1·68; Translucent.	—
Lapis-Lazuli	Deep blue with some mottling, often spangled with pyrites.	H.5·5; sp.gr. 2·75–2·9.	—
Opal	*Precious Opal:* rare. Good play of colour. White opal; black opal; Fire-opal (orange and red); Prase; green opal; Resin opal: yellow; Wood opal; like wood.	H.5–6·5; sp.gr. 1·44–1·47; Opalescent shimmer; is characteristic of opals.	Heat-sensitive.

name, type	*description, colours*	*hardness, fracture; sp.gr.; lustre, etc.*	*other characteristics*
Peridot	Gem form of olivine. Olive greens, yellow to brown-green.	H.6·5–7; sp.gr. 3·34. Transparent.	—
Quartz varieties:			
Aventurine quartz	Green, yellow or reddish brown with spangles of mica.	H.7; sp.gr. 2·65–2·66; opaque.	—
Cat's-eye	Yellow chatoyant.	Similar.	—
Tiger-eye	Golden-brown, striped chatoyant.	Similar.	—

CHAPTER FIVE

A Pebble Collection

You could simply place the flint nodules and pieces of rock and all the pebbles on shelves round the room. They make fine objets trouvé. But it is fun to make a collection, apart from making order out of creeping chaos. The passion for accumulating rocks can spread; chunks of stone fall out of drawers and stand in all the available spaces in cupboards, not to mention the cardboard boxes of them, stowed away in corners.

What sort of collection you make depends on the space you have available, and which side of the pebble hobby interests you most. Nothing could be more delightful than a comprehensive collection of beautiful water- and wind-worn pebbles. It is interesting to see variations on a single rock or mineral type, too, grouped together; a pebble or two, a piece of rough rock with one freshly broken face and weathered faces; or if it is a mineral pebble, large and small crystals of the species, and perhaps a polished example, and a stone worked and set in jewellery.

HOUSING THE COLLECTION

Where to keep your collection? If you are able to set aside a room for a workshop, then you can line the wall with lighted shelves or glass-fronted display cabinets. It is quite practical to keep the collection in the room you live in, if you choose a decorative piece of furniture to keep it in. Victorian small mahogany medicine chests or miniature chests of drawers do well for this purpose, and

FIG. 18 Storing a pebble collection

look attractive; so do little, old-fashioned display cupboards. The Victorians were so fond of collecting things themselves, they have left all sorts of pretty little sets of shelves and drawers for us to track down and use.

But modern office furniture can be stylish these days, and a bright, well-designed, vermilion-lacquered filing cabinet, provided that the drawers are two to three inches deep, would house a collection of medium-sized pebbles effectively.

The drawers would need to be lined with clean white paper, or light, plain material. Thin layers of inexpensive cotton wool look attractive, provided that you are not storing fragile crystals which will catch wisps from it.

KEEPING A RECORD

Apart from the storage cabinet, you will need to label each specimen on the less attractive surface with a number, which you can record in a card index or record book to show where the specimen was found, the date, and any other points of interest. For the labelling you could stick small paper labels, the sort prices are marked on, onto the pebble lightly, and write on the number. Or you could type the numbers on gummed paper and stick it to the stone. Marking the number directly onto the specimen is more doubtful. Supposing you change your mind and reorganise your collection with different numbering? It can be difficult to erase ink—from a porous stone, in particular.

Painting a small white patch on the specimen, and writing the number on it in ink is another method. This has the advantage that you can always paint a new layer of white over the old figure if you decide to change the number, and it will look neater. To do this, use a small tube of white paint and a small brush, and black waterproof Indian ink. These materials can be obtained at artist's material shops or large stationers' shops.

It isn't necessary to put the name of the stone on display beside it, although this has the advantage that any casual visitor looking at the collection has information immediately before him. It is necessary to record details of the collection beside the appropriate index numbers. Some people like to use label makers which emboss names on coloured plastic adhesive tape for labelling their collection, though the tape can be knocked off a glossy stone when it is being handled carelessly.

The record book or card index being prepared, and the labelling method decided on and materials gathered, the cabinet lined, clean and ready, you will need to decide what sort of a collection you want to make it. One of everything? Or all the permutations of each type grouped together?

Now is the moment when you will be glad you packaged the pebbles and labelled them when you found them. But, until you have organised the nucleus of the collection the way you want it, keep the packing labels with the specimens.

ARRANGEMENT AND CHOICE OF PEBBLES

Grouping the pebbles in order of the rock category they belong to —igneous, sedimentary or metamorphic—with mineral pebbles and other specimens kept apart, you can see what you have got. It is rather a slow operation to begin with, picking out the most representative and handsome stones. For example, suppose you have a number of breccia pebbles to choose from (the stones with angular fragments of mixed stones in their texture), which is the best to choose? It would need to be the one with the best demonstration of what breccia is, showing the even mixture and plain visibility of the fragments of stone inclusions. It should be an attractive stone too, in good shape in every way, without fractures and chips. If possible it should be a fair-sized representative of the species.

The process of selection and elimination begins. Be ruthless. Only include stones which contribute something, even if the difference is only one of colour. Each example should have something original to say in your collection, either by its quality as an example of its type, or its unique pattern, or the fine workmanship of the jeweller's faceting.

If space is very limited, it helps to plan the collection by making a list of all the things you want to get into it, and assessing how much space you can allot to each one. You can always expand the collection at a later date.

The size of the stones depends on the size of the storage shelves or drawers. If you have lots of space, then you can have some imposing pebbles, though others are bound to be small, in their nature. But smaller pebbles can go in a miniature chest of drawers, and still form a very interesting collection.

Knowing what you want will set you pebble hunting in earnest. A collection projects the study of rocks and stones into actuality, and makes everything written about them more comprehensible, apart from being interesting and a source of pleasure.

CHAPTER SIX

Polishing Pebbles

Some rocks will not polish. It is not in their nature. Chalk is an example of soft powdery rock which would disintegrate if you tried to polish it. Mudstone is another.

With slightly harder rocks, above hardness 3 in the scale of hardness, it is possible to begin to think of polishing them. Indeed, some even softer rocks can, with extreme care, be shaped and shone. Alabaster is one example; fossilised resin, amber, is another.

FIG. 19 Porphyry pebbles

Rocks can be polished by hand, or on a lapidary machine, or in one of the machines which has only come into popular use in recent years, a tumbler polisher. Whichever method you use, you will have the same materials to work with, either silicon carbide grits in various grades from coarse to fine, or diamond abrasives, and polishing agents used with hard felt, leather or wood buffs.

When a pebble or flat surface on a rock is polished the first stage is to grind away the rough areas, and any cracks on the surface, rather like planing wood smooth. A smooth, water-worn pebble won't need much attention at this stage, and soft rock is soon ground to shape. Soft rock would never stand up to the rough tumbling process in coarse grits which lasts for several weeks, day and night. It would be ground to a powder and never seen again. Though as your skill increases, you may begin to experiment with tumbler polishing rocks *slightly* softer than hardness 6.

Continuing the analogy with preparing wood for polishing, a series of abrasive stages follows the grinding stage, as a series of sandpapering processes follows planing wood, moving as each grade of abrasive has carried out its job of evening the surface still further, to the final silk-smooth surface which must be achieved before polishing.

Each method of polishing is particularly suitable for certain purposes; hand polishing if you wish to polish soft rocks or fragile minerals carefully, or specimens which are too large to go in your tumbler. Or, of course, if you do not own any polishing machinery, but would like to put a gloss on some of your pebbles.

Polishing on the lapidary machine means that you can control every stage, and cut down on the wear and loss of bulk to the stone. You can shape and polish stones as you choose, and treat each one individually, without the disadvantage of the very slow method of hand polishing when working with hard minerals.

Tumbling takes the effort out of polishing stones. You can deal with a load of about 3 lb. at a time in an average machine, and all the work you have to do is to change the grits and clean up the stones, and supervise the operation.

Some people content themselves with giving the stone a coat of picture varnish. This brings out the colour too, which is one of the main things polishing does for a stone. Giving it a high permanent gloss is the other.

I have even known one lady who produced a very good effect by shining her slate pebbles with some of the household liquid which claims to clean and polish the floor in one operation. Not having tested this method I wouldn't dare recommend it, though it is worth trying out on a few nondescript garden pebbles to see what happens!

HAND-POLISHING

For this method you will need three or four grades of silicon-carbide grit, a manufactured, silver-coloured abrasive grit, which is extremely hard. It is sold by mesh sizes, that is, the number given to the grit is the number of holes to the square inch in the mesh that it was graded by. You can use grit 60 or 80 mesh for the first process, but for a smooth, water-worn pebble, you can begin with 220 mesh grit. You will also need 400, and 600 silicon carbide grit; and some cerium or tin oxide polishing powder.

You will also need some pieces of wood about the size of medium shoe brush backs. Each piece of wood should be covered with hard cobbler's leather attached by strong adhesive such as epoxy resin. Alternatively hardboard can be glued to the wood and covered with a tightly-stretched layer of felt tacked to the back of the wood. A layer of tightly-stretched wash leather can also be added and tacked at the back. Do not use leather with dye in it, or when polishing porous pebbles the dye might leach into them and stain them.

To put the grit in, you will need some toughened glass basins. Medium-sized pudding basins work well, but don't use the cooking utensils from the kitchen. Keep a special basin for the purpose. You can pick up old dishes from the Oxfam shops or junk shops.

Have a bowl or bucket of water nearby for rinsing your hands as you work, and clean rags, and a kitchen paper towel roll. Keep handy a squeeze plastic bottle of the type used to contain washing up liquid, filled with clean water, and with the nozzle replaced. Wear an overall and rubber gloves, although there is no need to get in too much of a mess.

Coarse grind

Put a couple of tablespoonfuls of 220 grit into the bowl and dampen them with water, enough to make it a thick soupy mixture, so it does not cling in dry lumps to the stone, nor is it too runny. Swish the pebble round and round against the sides of the basin, moving it clockwise and anti-clockwise and grinding all the surfaces evenly. Rinse and dry the pebble every so often (on some paper from the kitchen roll) and inspect it through your magnifying glass, or closely under the light. Until you can see that all rough patches and scratched surfaces are removed and the entire surface is smooth and evenly ground, you will have to continue this process. Use the squeeze bottle to wash off the stone into the bucket.

With a pebble which was already worn smooth before you began, it should not take too long before you are ready to go on to the next stage. Don't give the pebble more grinding than it needs. A pebble of fine-textured slate won't need a lot of attention; a rough bit of quartz on the other hand will take an age to even out, so you can settle down to making it a long job, or doing it a little bit at a time every day until you are satisfied with the result.

Medium grind

This is the sanding stage. Wash everything with which you have used the coarse grit thoroughly so that not a single particle is anywhere near the work in hand. This scrupulous care has to be taken, because a grain of coarser grit can scratch a fine surface, and you may have to start over again, if this happens.

You could put the coarse grit aside, and use another basin—not only to avoid the kind of damage mentioned above, but also to be able to use the coarse grit again.

Use the same amount of grit as for the coarse grind, but this time 400 grit. Repeat the process, but spend rather more time on this stage, as you have got to be absolutely sure of getting rid of even fine, barely-visible scratches.

Fine grind

This is called finishing. Repeat the cleaning-up operations and the process of grinding, using 600 grit, making sure by close inspection that the surface of the pebble is silky smooth before you complete the grinding process.

Now comes the polishing stage. Before you begin, have a grand clear up. Put away all grits and cloths that have been used in the grinding process. Throw away any grit solutions you don't want to save—but not in the plumbing, or they might silt up the pipes. Use fresh water and cloths, and wash your hands before you commence.

Polishing

The reason why you use rather a hard buff for polishing a stone is that you need to be able to offer some resistance to the hardness of the material you are working on; to be able to set up some degree of friction.

Damp the buff and tip a little ceric oxide out into a dish, and add water to it to make a thick paste. Brush the paste onto the buff in a thick slurry. You can polish the stone with the buff or rub the buff on the stone. Either way this is a quiet process, and you can listen to the radio while you steadily polish without haste.

And lo! You have a gloriously polished pebble.

To polish a flat surface by hand

To polish a flat surface by hand the procedure is the same, except that you use a piece of plate glass or Perspex placed flat on the table. Tip the grits onto it, dampening them as before. Then press the flat surface of the stone onto it, and work it round and round, to and fro in figures of eight.

You can also make a large flat buff by mounting hard undyed hide leather on a disc of wood with a powerful adhesive. The stone is rubbed into the powder on it.

Using wet-and-dry paper or cloth

Another method of polishing pebbles by hand is to rub the stone with wet-and-dry diamond or silicon-carbide grit wet-and-dry paper or cloth. You follow the same sequence of grinding, sanding, finishing and extra-fine finishing as when using loose grit, omitting the first stage if you think that the last three grades of grit will be sufficient for a particularly smooth pebble.

The different grit sizes are bonded onto paper, or cloth, with the number printed on the back. Use the paper wet for stone polishing. It is available from some rock shops. See their catalogues.

Before beginning operations, get wet-and-dry paper ready: 60 or 100 grit, 220, 320 or 400 grit, and 500 or 600 grit (a choice of grit sizes is given in case stock is limited in your local shop). Wet-and-dry grit paper can be mounted on a wooden block over a thin layer of felt. Use strong adhesives, such as epoxy resin or tack on back

BEACH SHINGLE

A PAINTED PEBBLE

MINERAL CRYSTALS ETC.
(some in matrix)
1 Garnets in matrix 2 Gypsum selenite crystals 3 Blue John Fluorspar 4 A rock crystal (quartz) 5 Labradorite 6 Connemara marble 7 Topaz crystals in matrix 8 Botryoidal chalcedony cluster 9 Malachite (rough) 10 Branch of small quartz crystals 11 Turquoise in matrix (earthy variety) 12 Green fluorspar 13 Black tourmaline crystals in matrix 14 Tourmaline crystals (green and pink)

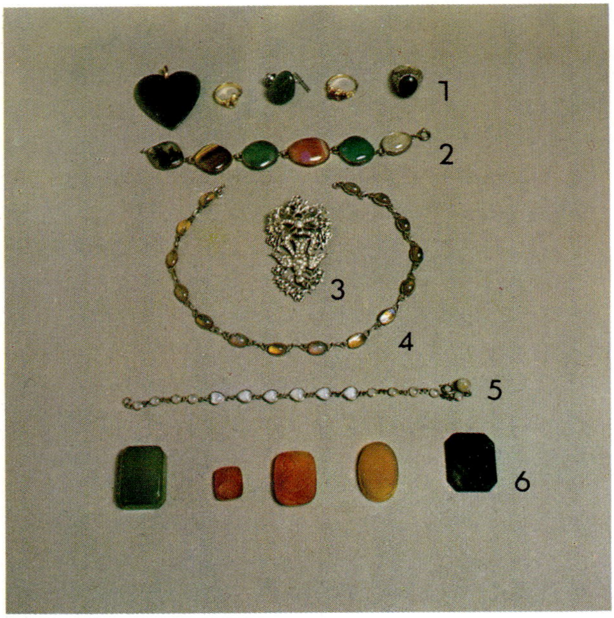

GEM STONES IN JEWELLERY

1 Various jewellery and jade heart 2 Bracelet of mixed agates and quartz 3 Rock crystal brooch 4 Fire opal necklace 5 Moonstone bracelet 6 Cabochons of chalcedony and green jade

AN INTARSIA AND MOSAIC PLAQUE

of wood. Put out the jar of ceric oxide polishing powder and polishing buffs like the ones recommended for polishing with loose grit. Have a bucket of clean water and a plastic washing-up liquid bottle full of water. Also paper towels and clean cloths, and your magnifying glass.

Both 60 or 100 grit, and 220 grit paper can be used on a buff flat mounted on the table, or held in your hand like a cloth. You can also use the paper or cloth like a duster. Keep the paper wet and clean while you are working by squeezing water over it.

Next, rub the pebble evenly all over, using wet-and-dry paper until you complete each stage, inspecting carefully before moving on, and washing everything each time you change over, especially for polishing. Polishing is carried out as before, on a perfectly smooth surface.

LAPIDARY EQUIPMENT

Unless you are able to use lapidary equipment regularly at a club or class, sooner or later you will probably hanker to own your personal machine, which would enable you to make jewellery and other ornamental items at home. The machine, if carefully chosen, should enable you to saw and trim rock, grind and shape it, and polish it. With the right accessories you should also be able to adapt it for faceting (grinding flat ornamental faces on a gem to reflect light, as for diamonds) and if a sphere-making accessory is available to attach to the machine, you can also use it for making beads.

An all-purpose lapidary machine will probably be adequate for average hobby requirements, especially if there isn't too much space to set aside for a workshop. You may prefer to have individual machines for each process, such as a trimming and slabbing saw, and a lapping machine for polishing flat slabs as well as powered grinding, sanding and polishing machines. It depends on your enthusiasm for the hobby, and the scope and extent of your activity.

As individual larger machines have to be housed, powered and maintained separately, probably they would appeal more to club or commercial users, although for anyone, they undoubtedly have the advantage of greater facility and scope in use, unless your all-purpose machine is to be a fairly large version.

At present, a much more limited range of small equipment is available in Britain than in the United States, but there are, nevertheless, some excellent modest-sized, all-purpose machines on the market. When choosing yours you need to take many considerations into account, including what range of accessories the machine has available. You can always add them later, one at a time.

However, before you invest in your own equipment, you might like to go to a class or join a club, so you can use the machines there, with experienced help. This will make it much clearer to you how a lapidary machine works, and which model you would eventually prefer, knowing now what you would expect from it.

If this is not possible, write to the rock and lapidary equipment suppliers and ask for their catalogues, so you can compare machines. New models are always coming on the market, and it would be maddening to acquire a machine, and then discover that there is another more versatile model for about the same price, which you would have preferred.

The addresses of some leading firms are given at the end of this book, but you might find other addresses from advertisements in a lapidary magazine. Try to visit one or two shops, too, before you finally decide on your equipment.

Your basic requirements

These would probably be a lapidary machine powered by an electric motor, with a coarse-grinding, silicon-carbide wheel, and finer grade sanding wheels called scratch wheels, or a rubber disc you can use in conjunction with silicon carbide or diamond grit wet-and-dry paper discs. You would also need a polishing buff, and one of hard felt is fine to begin with. Buffs of leather and plywood are also available. Polishing powder is needed to use with the buff. The other item is a diamond saw blade, which is likely to be rather expensive—at least £5, even for the smallest size. If you

PLATE 1 Above: a Robilt combination machine with saw, grinding wheel, etc.

PLATE 2 Below: feeding a piece of clamped rocks into a diamond saw blade.

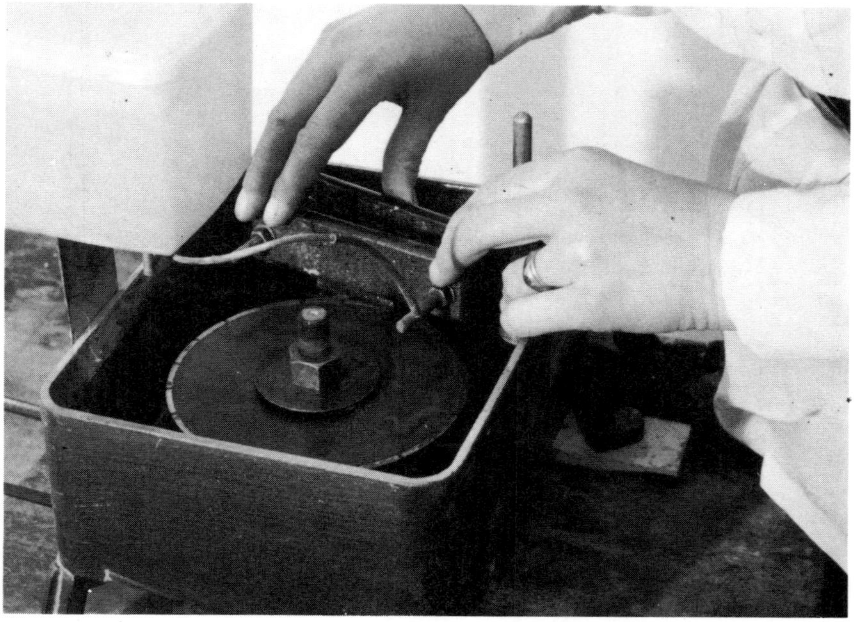

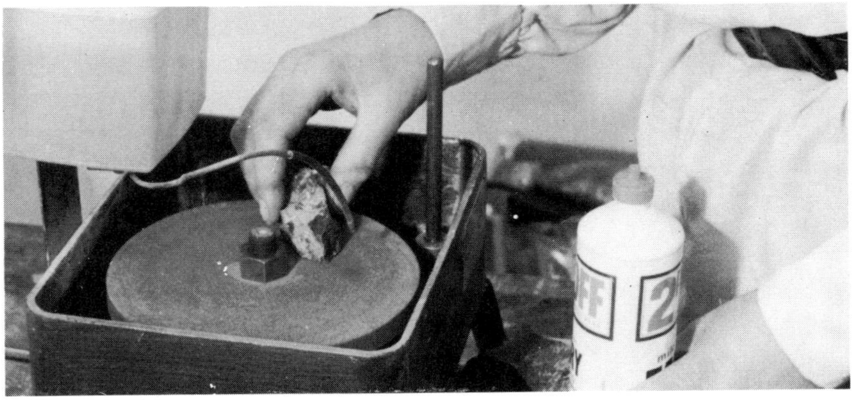

PLATE 3 Grinding a piece of serpentine rock on a silicon carbide wheel.

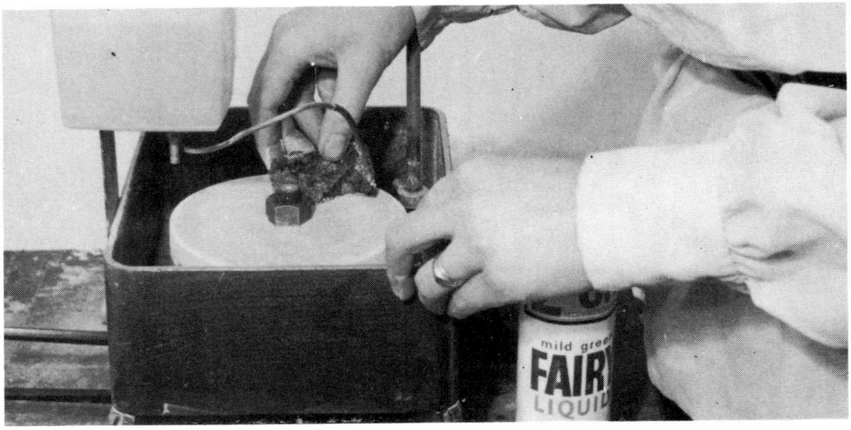

PLATE 4 Polishing serpentine rock on a felt buff.

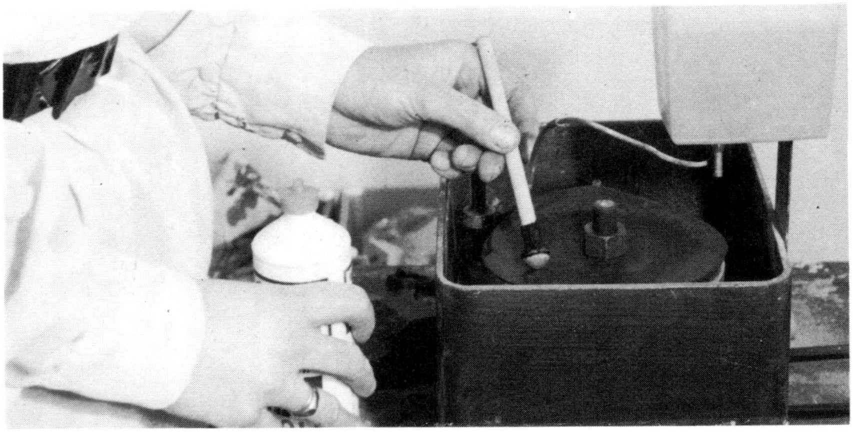

PLATE 5 Sanding a cabochon on wet-and-dry paper disc.

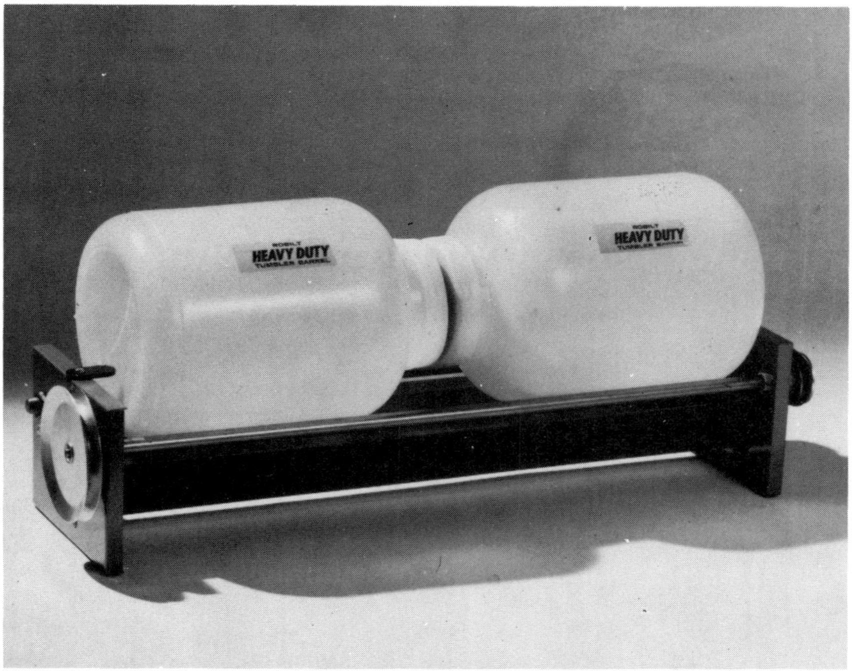

PLATE 6 Above: a Robilt tumbling polishing machine.

PLATE 7 Below: an inexpensive 'Tumble Master' machine by M. L. Beach (Products) Ltd.

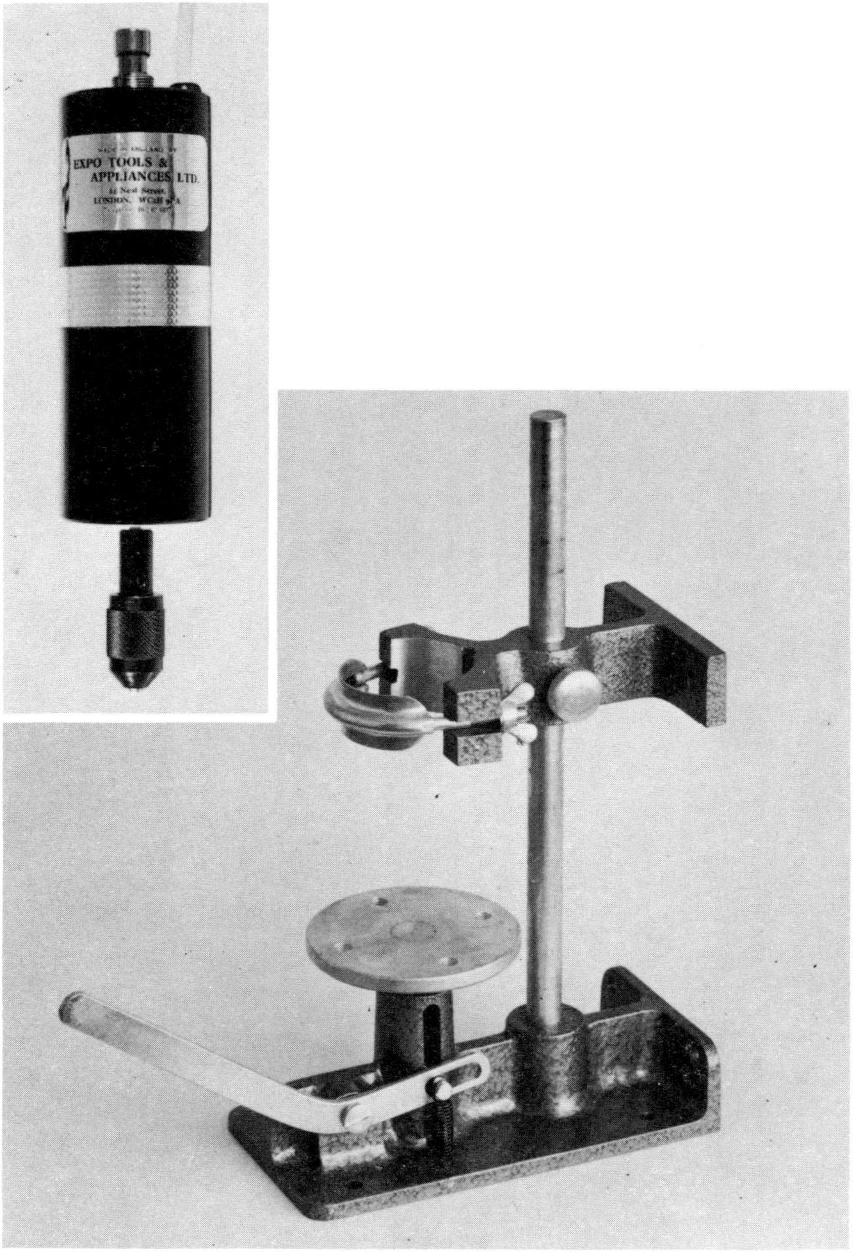

PLATES 8 and 9 An Expo 'Titan' diamond drill and drill stand.

can possibly buy one, do so. There are many ways of working with stone for which you will wish you had one.

Accessories

Diamond laps, usually copper impregnated with diamond grit are available for fine work, but for ordinary pebbles and stones are not a necessity. The faceting and sphere-making attachments have been mentioned already. One machine has a precision faceting attachment which costs as much as the machine and its basic accessories complete! A more modestly priced faceting attachment is also obtainable for this machine.

If you choose a machine with a horizontal wheel, that is, one with a vertical shaft, a useful accessory is a cast-iron or steel lap—a flat plate, fitted on the shaft, which covers the entire top of the machine. With this lap you can polish flat surfaces of fair-sized pieces of rock, and this is very helpful if you want to polish large inlay squares of marble or other stone for table tops, book ends, etc. Various types of polishing buff are other possible attachments.

Horizontal or vertical wheels?

This is one of the most important questions you will need to consider. Some machines are built on a horizontal shaft, either with the various grinding and sanding wheels and buffs fixed beside each other in order along the shaft, or with the wheel fixed to the side of the machine, so the various laps can be changed for different jobs. A horizontal shaft offers certain advantages when sawing and cabochon making (the smooth shaping of stones for jewellery), and this especially applies to sawing. A larger, all-purpose machine usually has the saw fixed at one side, the blade in a vertical position revolving through a coolant tank, and has a sludge pan that is easily cleaned. This is definitely a more efficient and accessible position for the saw blade and process. It means that you can feed the rock into the blade and see what you are doing. On the other hand, with a machine with vertical wheels, you would need a separate machine for flat lapping.

When you are choosing a machine with a vertical shaft, and horizontal grinding wheels, look at the arrangements for sawing

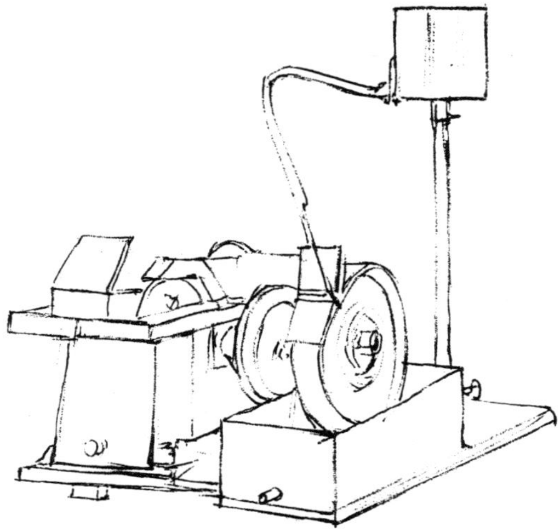

FIG. 20 A combination lapidary machine

very carefully. With this type of machine, the stone to be cut has to be fitted into a clamp at the side, which has an arm with adjustable screw plates between which the stone is fitted. The stone protrudes below the plates, level with the saw blade, and when the machine

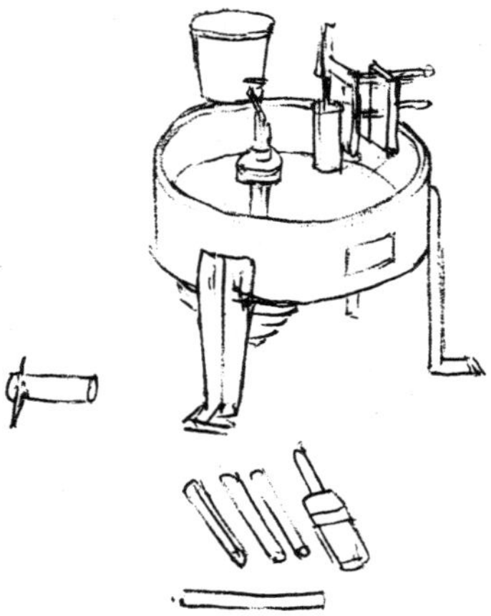

FIG. 21 A Robilt Gem-maker

is set in motion, it is fed into the saw from the side. If the space between the side of the tub and the saw blade is small, you will be restricted to sawing small stones. You would, in any case, need a separate trim-and-slab saw for working on large chunks of rock.

Look at the arrangements for feeding the rock into the saw. The screw clamp should be strong and efficient, and if possible have some guide method for the repeated cutting of slabs of equal thickness from the same piece of stone.

The saw blade

The diamond saw blade has very fine-mesh diamond grit embedded in the edge, either sintered, which is with grit embedded all round the rim, or notched, where the diamond grit is embedded in notches. Normally they are safe to use, but there are very sharp versions for specialised lapidary work.

The machine will need a strong, easily-cleaned tub, a coolant tank for holding water or coolant oil, and a sump for waste; preferably, it should have a simple shape for easy cleaning, for lapidary work can produce a lot of messy sludge.

The motor

The machine may be motorised already, which after taking all other points into consideration, will save you some trouble. When you buy the machine the firm will probably be able to recommend a motor supplier, or perhaps be able to supply you from stock.

New motors are expensive, and could cost a third to half as much as your lapidary machine, so it is worth while tracking down a second-hand motor of the recommended horsepower, which is usually $\frac{1}{3}$ or $\frac{1}{4}$ horsepower. Before you set off to the motor supplier (usually an engineering retail firm), write down the size of the pulley wheel on your lapidary machine (ask your supplier what the measurement is if you don't know how to assess it yourself). At the same time, ask him what size and width pulley belt is recommended, and what size, type and width of pulley wheel is recommended for the motor. Without all this information you may waste your journey.

Setting up the machine at home

To do this, you will need a strong, stable bench or table. A small, old-fashioned, white-wood kitchen table is ideal. You can have the machine screwed down at one end, and the motor at the other, the correct distance away; or, alternatively, on a shelf built below, if your machine has a horizontal shaft. The pulley wheel has to be aligned with the pulley on the motor, and the motor should be shielded from flying water or liquid.

The electric socket, and all electric wiring, must be fully earthed. The motor, if not fully insulated against water penetration, must be protected by a splash guard. And, when choosing a machine for a club, check whether two or three people can work on it at the same time. Some machines are designed with this in mind.

The tools used for jewellery making are described in chapter 7, including a description of a diamond drill and how to use one for boring holes through pebbles to fit jump rings.

SAWING

You may want to saw some pebbles in half, especially if you think you have found a potato stone, or geode, which could have a hollow inside lined with crystals. Or perhaps you may want to slice a suitable pebble into thin slabs, to polish and set as jewellery.

The other way you would be likely to use the saw would be for sawing through chunks of rock (which would need to be not too large for a hobby-sized machine), in order to make flat slabs to polish for inlay work, or for ornaments. Trimming shapes is another use for your rock saw.

Pebbles can be tricky to saw because of their smooth shape and curved surface, which makes it difficult to fix them securely into the rock clamp. Various types of vices are fitted to superior saw machines, and these make for firmer fixing.

The process by which a diamond saw blade cuts through rock is by grinding it. Diamond (the hardest mineral of all) will, by its great hardness, grind away softer rock, and this is how the saw works. The diamond saw blade is fitted on the shaft (unless the blade is permanently fitted on a trim-and-slab machine), the lubricating liquids, which are vitally important to the operation to prevent wear of the saw blade, are checked, and the motor is switched on.

The saw blade rotates very rapidly, and the diamond grit embedded in the rim of the saw grinds away slowly and steadily as the stone is being fed into it, until the entire piece of rock or pebble has been cut through. The width of the cut into the stone will be equivalent to the thickness of the saw blade.

Some diamond saw blades are intended to run in one direction only; others can be reversed. If yours is the former kind, before you start work paint *'this side up'* on the upper side, or you may forget and reverse the blade next time you use it, and so spoil it. Refer to the maker's instructions to find out what is recommended for your saw blade.

Lubricating the saw

It is essential to lubricate and cool your saw blade continuously while it is at work. The lubricating liquid, often referred to as the 'coolant', keeps the blade cool at the point of friction between the spinning saw blade and the rock that is being cut. At high speed the heat set up could warp and damage the saw blade and wear it out rapidly if the point of friction were not generously lubricated.

Water is a satisfactory lubricant for various stone-cutting processes, such as grinding, sanding and polishing. But a suitable form of oil is also recommended, particularly when sawing. Oils used with stone-cutting equipment are usually the light cutting oils used in engineering. For domestic use, for various reasons, cutting oils which emulsify in water (called soluble oils), are preferable. The water reduces their flammability in comparison to neat oil, as they are usually used diluted twenty to one hundred parts of water to one part of oil, depending on the make of oil that you are using. Suitable soluble oils include Dromus B, made by Shell-Mex, and Energol SB4, made by B.P. Another, Miracol 80, was developed especially for use by the stone-cutting industry, and is manufactured by Burmah-Castrol Ltd.

There is some difficulty at present in obtaining these oils in small quantities. Usually a gallon is the smallest amount that you can buy, though this will keep for as long as you want. Smaller quantities can be obtained from the sources mentioned at the end of this book, in the address list, which also includes names of retail suppliers of one-gallon drums.

Paraffin is another oil which is recommended as a coolant, but the powerful smell and highly inflammable properties of this liquid make it unsatisfactory for use indoors, unless it is insulated in a coolant tank through which the saw blade runs. An open drip of paraffin, and squirted paraffin from a plastic washing-up-liquid bottle could be a fire hazard. Extreme care should be taken with it and any oil, and fire safety precautions observed, if these are used. They are *dangerous!*

Preparing to saw

Before beginning work, you should organise the coolant liquids and a pail for the waste from the tub, unless your machine is fitted with a removable sludge receptacle. The sludge should be thrown away on earth, not down the domestic plumbing, or it could block the pipes. If you wish to cut a slab from a stone of a particular thickness, now is the time for marking it with an aluminium scribing pencil (see page 123).

Clamp the pebble

You will soon be able to judge what size stone your machine can deal with. The pebble or rock piece must be fitted firmly in the clamp, so that it will not move when the pressure of the saw is bearing rigorously into it. It may need to be wedged if it is very slippery or an awkward shape, and for this purpose you can use little pieces of wood or hardboard. You won't want to have to hold the pebble in position in the clamp while sawing it, because it is not held tightly enough by the screws; and, if it does slip about, it could be damaged against the saw.

Slipping about is particularly associated with a saw blade mounted horizontally on a vertical shaft. It is much easier to feed a stone along a level table towards a vertical saw blade, although here too, clamping the stone has to be done with care.

The hardness of the stone you are cutting will also affect the time that it takes to cut it through. A piece of bloodstone, hardness 6·5, will take far longer to cut through than a piece of serpentine rock, which has a hardness grade of 3–4.

The Sawing

Having checked that the coolant tank or trough is replenished with oil or water, you should, if your machine has a tank which drips the liquid onto the horizontal wheel, open the tap. You can supplement this type of coolant drip with a plastic squeeze bottle filled with the *same* lubricant as is in the tank. If possible, direct the lubricant from this bottle to the point of contact between the rock and the saw blade, so it lubricates the under as well as the upper side. This will help to prolong the life of the saw blade.

When the pebble or rock piece is clamped in position, and the motor is switched on, make sure that the pebble is held away from the blade until you can control the rate of feeding it into the saw. When you feed the pebble gently in, the blade will begin to grind into it. Remember that this cut, a groove in the pebble, must be continuously lubricated during sawing.

Do not try to get quicker results by applying extra pressure on the clamp to force the rock into the saw. This way, you may overheat and buckle the blade. Feed the rock into the saw's grinding edge with just enough pressure to help the process along, but without haste or impatience. The power of the diamond grit in the blade is only as great as its relative hardness to the material it is cutting. Slow down if the blade begins to wobble, and again, always remember to keep the blade well lubricated.

Finishing the cut

When the pebble is almost cut through, if you don't anticipate it, the last fraction may snap abruptly, leaving a sharp point of rock sticking out which could swerve across the saw blade and damage it. If this happens, draw the clamped rock away from the blade smartly and instantly. You will need to acquire the knack of holding the blade more lightly into the saw as this moment approaches, in order to make a perfect cut. Otherwise, snap the point off afterwards with pliers.

Cutting parallel slabs will not be difficult if your saw clamp supplies you with some method of measuring a second cut without having to move the stone. The more superior machines are built

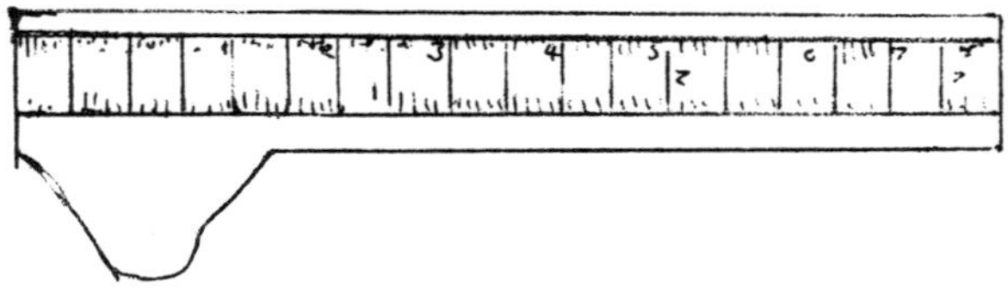

FIG. 22 Sliding gauge for measuring slabs of rock

with an automatic cross-feed clamp which takes care of this operation, but with the ordinary machines, you may have to work it out for yourself, with the help, possibly, of a slab-measuring gauge, obtainable from some rock shops.

One method is to leave enough rock sticking out of the clamp to cut more than one slab out of it, but this has the disadvantage of giving less stability, because the stone is not so firmly gripped in the clamp. Not too much stone should protrude from the clamp beyond the line along which you intend to cut, and you should make sure that you have reset the stone for an exactly parallel second cut to the first before proceeding with a complete cut, or you may have a lop-sided slab at the finish.

POLISHING A FLAT SURFACE

A rock specimen can be enhanced by polishing one side to intensify the colours and contrast them with the natural fractured and weathered surfaces. Flat surfaces are polished for inlay work, such as intarsia mosaic pictures (see page 145), or when a form has a slab shape.

Preparation

Assemble—or prepare—the grinding and sanding wheels of your lapidary machine, to include the coarse-grit, silicon-carbide grinding wheel, 100 or 120 mesh grit, another of 220 mesh grit, and others 320 or 400 mesh, and 500 or 600 mesh grit. Alternatively, apart from the coarse grinding wheel, you can use wet and dry silicon carbide or diamond grit paper discs fitted on rubber discs.

You will also need, later, a felt buff, which should be kept in a polythene bag until used, to avoid contamination by any particles of coarse grit that could be about. You will also need polishing powder, which can be tin oxide or cerium oxide powder. Replenish the coolant tank with the lubricant that you prefer (a special water-emulsifying oil, or water) and prepare a plastic squeeze bottle of coolant, if your machine has a coolant tank which drips onto a horizontal wheel, which may need supplementing from time to time. Or follow the maker's recommendation for your machine.

Have your magnifying glass ready, and a bucket of clean water and kitchen paper towel handy, so that you can rinse and dry the stone, and inspect it during operations. Always change the water carefully when moving to a finer grit, so that you avoid the possibility of contact with coarse grit, and work in a good light.

The Coarse grind

This is intended to remove the roughness and any scratches on the surface to be polished, and to make the surface even. Use the silicon carbide 100 or 120 grit wheel.

If your coolant tank is above the wheel, open the drip tap, and let some lubricant drip onto the grinding wheel for a few seconds before switching on the electric current. When the motor is switched on, it will spin the grinding wheel rapidly and spread the lubricant drip, which may need to be supplemented from time to time. This applies particularly to horizontal wheels.

Press the surface to be polished onto the wheel, and move the rock to and fro, clockwise and anticlockwise, across the whole surface of the grinding wheel. Take care to avoid knocking the stone against the shaft and chipping it, or grazing it along the edge of the wheel, and causing a groove or scratch to form.

Every so often, rinse the rock and dry it. Inspect it through the magnifying glass if necessary. When the surface looks flat and smooth all over, and any deep scratches have been eliminated, wash and clean away all traces of the coarse grind, because of possible subsequent damage by coarse grit particles, and move on to the next stage.

Sanding : medium grind

This is a finer sanding process using the 220 grit wheel, or wet-and-dry paper disc on a rubber disc. Use the same procedure as for the coarse grind, this time dwelling with extra care on the removal of any remaining shallow scratches. Even small hair scratches will show up through the final polish and spoil its appearance. Again, when you are satisfied that you have made the surface as smooth as possible at this stage, clean and wash everything before moving on to the next stage.

Sanding : fine grind

Use the 320 or 400 grit wheel or disc, and rotate the polishing surface of the rock as for the two earlier stages, inspecting the surface at intervals very critically through the magnifying glass, in order to detect any remaining blemishes. If any are still to be seen, continue smoothing the rock in the same figure-of-eight, to-and-fro movement, until you are satisfied that the surface being ground is completely smooth. Do this thoroughly before going onto the next important stage, for finally getting rid of scratches in the stone. Clean and wash everything carefully before proceeding any further.

Sanding : extra-fine grind

This is a refining process to bring the rock surface to a state of almost-polished, silk smoothness. Use the 500 or 600 grit wheel or wet-and-dry paper disc, and repeat the action of smoothing the rock on the moving wheel, as with the first three stages. When, having looked carefully at the surface through the magnifying glass, you are content that it is perfectly smooth and free from even minute blemishes, you can go on to the polishing stage. Some people like to pre-polish using levigated (very fine) alumina powder, the equivalent of about 1000 mesh grit, on a separate buff, but this is not essential

Polishing

Clear away all the accessories and materials that you used in the grinding and sanding processes, and clean and wash everything thoroughly. Only when you are satisfied that not a grain of silicon carbide is lingering about should you take the felt buff out of its polythene bag and fit it on the wheel—if yours is an all-purpose machine, that is. On a fixed machine, the buff will remain in position throughout. Mix a couple of tablespoonfuls of cerium- or tin-oxide powder into a thick paste, and apply a layer to the dampened felt buff, using a brush.

Rub the polishing surface of the rock into the paste, as the wheel spins, to and fro and round across the width, rinsing in clean water and blotting dry before inspecting the stone through the magnifying glass to see how the polish is developing. Now you will appreciate the reason for getting rid of the scratches so assiduously at the earlier stages, when you see the gloss appearing on the rock. Try to polish long enough to work up good friction between the buff and the stone, as this helps the process.

It will only take a few minutes to achieve a magical gloss on a well-prepared stone. But, if the polish doesn't seem satisfactory, and you are sure that you have chosen a suitable stone for polishing, then you can go back to an earlier stage and repeat the process all over again. At the second attempt, you might get better results.

The last stage is clearing up. Make sure that the grit wheels are rinsed clean and standing on their sides until they dry out (they mustn't be left water-logged on the machine). If yours is a machine with horizontal wheels you will need to remove and rinse the felt buff, dry it, and then put it away in its polythene bag.

In addition, the screws and metal supporting lap should be taken from the shaft and washed and dried thoroughly, to avoid rust. And rinse the tub and machine thoroughly to remove the sludge and grit and any other waste that has accumulated. Throw the waste away—but not in the plumbing.

Polishing a large, curved pebble on the wheel

The secret of polishing an extra-large, curved pebble is to keep it imperceptibly moving over and over, always at a changing, slightly oblique angle, so that no flat surfaces have the slightest chance of developing. Try to keep the shape true to its natural curve by looking at it in its profiles at eye level, and by looking at its shape from above, with and without the magnifying glass. The grinding and polishing process follows the same routine as for polishing a flat surface. But if the pebble is already water-worn smooth, you may decide to eliminate the 100 or 120 coarse grind.

USING A FLAT LAP

A combination lapidary machine with a vertical shaft, on which the various grinding and sanding wheels, saw and buffs are interchangeable, has the advantage that a flat lap can be fitted on the shaft, as an accessory. When you are buying a machine it is wise to enquire whether a lap is available to fit it, though it is sometimes possible to have one with a different size screw adapted to fit your machine. Alternatively, lapping machines can be bought separately including a vibrating lapping machine which shakes under the stones being ground, and does the work of smoothing them for you.

The lap is a circular plate, usually made of cast iron or steel, with a solid surface. The shaft does not protrude at the centre, and this means that the entire surface can be used for polishing larger pieces of rock.

The grinding and sanding process is the same as for polishing a flat surface by hand, using loose grits in different grades made into a slurry with water, spread on the surface of the lap. Silicon-carbide grit is used, starting with 60 to 80, if the rock is large and rough, or 200 mesh grit if it is relatively smooth. You proceed by stages, through 400 grit to the pre-polishing stage, using ultra-fine grit powder.

Polishing will need to be done on the usual buff surface, using a polishing agent such as cerium or tin-oxide powder. The whole surface of the lap should be used to prevent indented grooves forming where the rock is rubbed habitually, and destroying the level surface.

Once more, it is vital that all the equipment, the lap, and your hands are cleaned thoroughly between stages, to avoid any contamination by coarse grit at the fine polishing stages. Grits could be kept strictly apart too, and anything used in contact with them.

PEBBLES INTO CABOCHONS

A cabochon is one of the simplest forms in jewellery, a smoothly-polished shape without facets, that is, flat surfaces. Cabochons can be square- or oblong-shaped, or, commonly, have a round- or oval-domed crown, or even be in the shape of hearts or crosses. The underside is flat with a narrow bevel around the edge. You don't have to be a lapidary to be able to make one, and, although gem stones are normally used, it is quite possible to produce beautiful and interesting cabochons out of suitable pebbles and rock fragments.

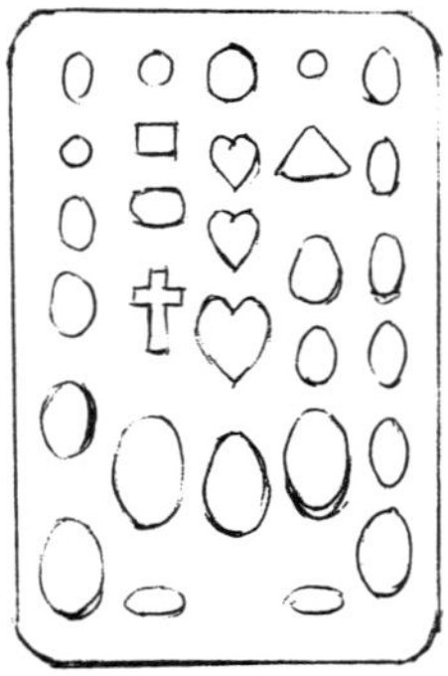

FIG. 23 Template for cabochons

Rock shops supply templates of cabochon shapes, sheets of plastic or metal with the shapes cut out so you can stencil them on the stone. The advantage of using a template is, as well as giving you a perfect outline, on a correctly designed one, the shape fits a standard-sized jewellery mount.

An attractive-looking stone which polishes well is the type to choose, preferably a mineral pebble. Chalcedony and the agates, with their waxy lustre are suited to the curved, cabochon form—especially as they are sometimes banded and marked in a way that can be included to enhance the finished stone.

Water-worn pebbles of suitable stone supply you with a half-formed cabochon already. If you select the stone carefully, you can follow the shape it had when you picked it up on the beach. But when you are selecting pebbles for cabochons, consider what you are likely to use them for. If you intend to mount them as jewellery, remember which size you like to wear. Fairly large, dramatic stones may be practical to wear on your fingers, but would be uncomfortably heavy set in earrings.

The materials for making a cabochon

The lapidary machine is used, although pebbles can be made into cabochons by hand, as well. If you want to make a large number they can be ground into shape first and polished in a tumbler polisher. Then, they are called 'preforms'.

You will require the same range of grits and polishing equipment as you used for polishing a flat surface: the coarse silicon-carbide grinding wheel of 100 or 120 grit, and other wheels or wet-and-dry paper, either silicon-carbide or diamond grit discs used with a sponge rubber lap, in 220, 320 or 400, 500 or 600 mesh sizes. Also, you will need a hard felt buff and some ceric-oxide or tin-oxide polishing powder.

Top up the coolant tank, and, if you are using water as the coolant, have some rinsing water and a kitchen paper roll by. And remember that your magnifying glass and pocket magnifier, giving larger magnification, are particularly necessary when making cabochons.

Dopping

To grind the small pebble while you hold it in your fingers would be a clumsy operation, so the pebble is usually mounted on a stick to enable you to manipulate it more easily. The shaping is done while you twirl and move the stone, cemented to the stick, over the surface of the spinning lapidary wheel. A cabochon mounted like this is 'dopped', and the stick is called a 'dop stick'. The black wax used to cement the pebble to the stick is called 'dopping wax'. So you will need some dopping sticks and dopping wax, as well as some old cake tins or meat tins and an old, small saucepan. The use of these will be explained later. You can obtain dopping wax from a rock shop, and dop sticks too, although you can easily make these yourself. The most suitable material to use for the sticks is wood dowelling, up to about half-an-inch diameter. You can make larger cabochons than this, but the larger ones can be worked with them held in your hand. The sticks should be cut in 4 in. to 6 in. lengths.

When you are dopping your sticks, putting the wax on the ends,

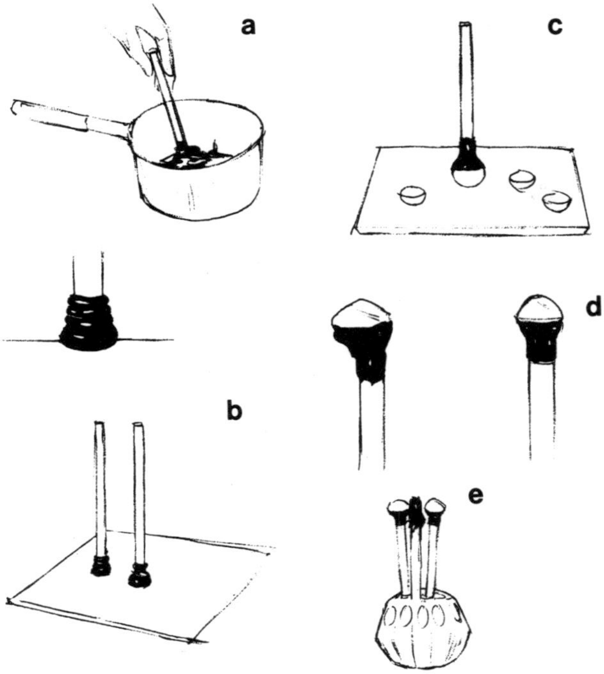

FIG. 24(a) Twisting wax around a dop stick (b) Waxed dop sticks cooling on a tray (c) Pressing dop stick on a hot stone (d) Cabochons: wrongly mounted (left), correctly mounted (right) (e) Dop sticks in a flower stand

you will need something to hold them upright. You could make a stand by drilling large holes in a block of wood. But a rose flower holder, the type that you put the stalks of flowers in, to stand in a vase, is ideal for the purpose. If you haven't got an old one at home, you can usually buy one cheaply in a large supermarket.

But before you dop the sticks and cement the stones on top of them, it will be necessary to flatten one side of the pebbles, the reverse side to the more attractive surface you wish to polish. Grind the flat side on the grinding wheel, using the 100 or 120 grit wheel, followed by refining the surface on the 220 grit wheel. Later you make a narrow bevel all round the edge. This is a flat-angled grind which eliminates the sharp edge and helps to prevent the stone from crumbling away in the grinding process.

When dopping the sticks, you should put a couple of cakes of dopping wax into a saucepan, and put it on the kitchen stove to melt them over a low heat. Have a bowl of cold water nearby to dip your fingers in, should they come into contact with uncomfortably hot wax. The wax should be hot, but not overheated.

When the wax has melted, take the first stick and dip it into the hot wax until the end is covered thickly to the depth of about half an inch. Take the stick out, twirling it to spread the wax round the end evenly, then turn it firmly, wax end down, onto the cake tin. The wax will broaden out and settle round the end.

After they have thoroughly cooled, the dop sticks can be stood in the rose holder for some time to harden.

Meanwhile, place the shape you have chosen in the template over the flat area of the reverse side of the pebble (choose one which fits the shape of the pebble, preferably), and draw its outline on the pebble with the aluminium wire or scribing pencil. Afterwards, grind a small bevel round the edge outside the outline.

To bond the pebble to the stick firmly enough for it to stand up to the grinding and sanding processes, without coming off the wax, both the pebble and the wax have to be at the right heat at the moment of cementing together. Except for heat-sensitive stones, such as opals, which need to be insulated from direct heat, on a tray of sand, or an asbestos mat, and heated gently even so, ordinary pebbles can be heated slowly in a very low oven in one of the cake tins, or on top of the stove in a thicker pan, again very slowly (flat side up).

When in your estimation the pebbles are hot enough, test one of them by dropping a fragment of wax on the surface, the way you test a hot iron. If the wax melts the stone is ready. Take a dopped stick. Work quickly and with care. Hold the wax end over the heat and, at the moment when the wax is hot and melting, plunge the stick firmly onto the flat side of the pebble.

Lift the stick up with the pebble on it, avoiding getting any of the wax on the rounded upper part, or it will get in the way of your grinding processes eventually, and swiftly mould the wax up under the pebble, centring the pebble over the stick as you do so. When the pebble is firmly seated and bonded, place the stick in the stand or flower holder for the wax to harden. Repeat the process with the other pebbles.

You can re-use the wax, so leave it in the saucepan. It is a good idea to prepare the dop sticks in a leisure moment, so you have some ready when you want to make cabochon.

Work on the cabochon

When you begin working, the pebble may annoyingly fall off the wax. This may be because the wax cooled slightly at the moment of bonding—in which case, there is nothing for it but to try again. But, if the pebble holds, you shape the cabochon on the grinding wheel, using 100 or 120 grit silicon carbide, holding the stick between your fingers and thumb, and moving the stone over and over evenly without resting on one spot for an instant, or a flat surface might form. Flat areas can form easily, especially on top of the stone, and they are very difficult to eradicate afterwards. But, if you tilt the stone slightly, all the time that you are grinding, and even when dealing with the top, you will avoid this. If a flat area does form, in spite of your efforts, you will need to reduce the height of the cabochon a little, in order to build up a curved shape again.

Not only do you have to keep the crown of the cabochon evenly shaped and curved, you also need to look at it critically from above to see its shape is even, and in profile, at eye level. When making a cabochon from a pebble it is best to let the shape of the pebble originally be your guide, while keeping to the template outline.

Stop often, clean the cabochon, and inspect it through a magnifying glass. Then you can decide that 'its a little too high at that end', or that 'one end is more rounded than the other. I'd better grind the other end to a slightly more rounded shape, to match'.

When you are satisfied that the shape is symmetrical, the surface smoothly ground, and all roughness and scratches eliminated, you can wash everything, clean the stone, and change wheels to the silicon-carbide 220 grit wheel, or to a similar-mesh, wet-and-dry paper disc, and proceed to the sanding process.

The secret of keeping the cabochon in a balanced shape, not lop-sided, and not low at one end and too high at the other, is constant surveillance and inspections, and extremely careful working on the grinding and sanding wheels.

After the sanding process, move on to the finishing process, using the 320 or 400 grit wheel or wet-and-dry paper, and after this, to the 500 or 600 grit extra-fine finishing process. Always clean up between processes, so that coarse silicon-carbide grit doesn't stray into a refining process and ruin it, by scratching the surface.

Polishing

When the pebble-cabochon is as smooth as glass, and with not a mark on it, even with the most hyper-critical inspection, then you can polish it. Fit the felt buff on the machine, and make a paste of polishing powder, a thick slurry, which you spread on the buff with a clean brush.

Polish the cabochon with good friction, and soon, as always on a well-prepared stone, a glorious gloss will be seen. To remove the polished pebble from the wax, place it in the freezer compartment of the refrigerator briefly, then slip it off the wax with a knife.

A CABOCHON FROM A FLAT STONE

It is possible to make a cabochon from a flat small piece of pebble, or a piece cut from suitable larger rock. To do this, you will need to use the same materials you used for dopping and polishing the pebble cabochon, plus some more items, such as the diamond saw blade of your lapidary machine, and pliers. To draw the shape, place the template over the best part of the stone, looking for the most flawless part, and the prettiest, combined. After the outline has been drawn on the stone, to take in this part, the stone needs to be trimmed to shape. If the cabochon shape you chose covers most of the area of the stone, you simply have to grind it to shape on the silicon carbide 100 or 120 grinding wheel, to within a short distance of the outline.

If the stone surrounding the outline is too large to do this con-

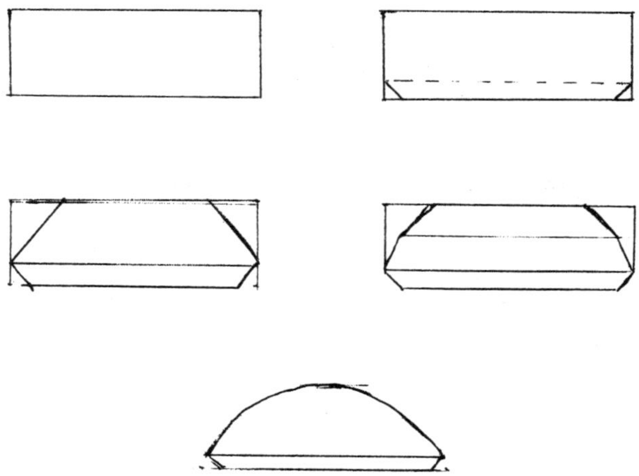

FIG. 25 Making a cabochon from a flat stone

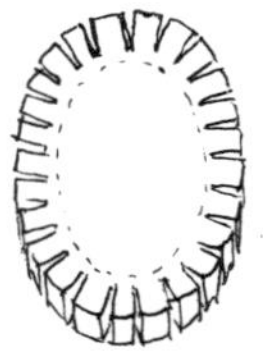

FIG. 26 Saw cuts for shaping a flat stone for a cabochon

veniently, then you can make short vertical cuts into the side on the saw, to within a margin of the outline, and break off the resulting notched teeth with a pair of pliers. This done, it will be necessary to grind the sides of the now roughly-shaped stone to shape, and give them a smooth surface to work from.

Assemble all the materials for the grinding, sanding and polishing processes, and all the dopping materials, as for making a pebble cabochon, and begin by making a narrow bevel round the base edge of the stone. Then make a second bevel, sloping at an angle towards the centre of the stone, from the lower to the upper edge of the sides. Make a third steeper angled bevel, round the upper edge of the stone, towards the top centre, and you will have formed an incipient dome.

If you wish to make a flat square cabochon, you deeply bevel the upper edge, and afterwards form a curve from the bevel, by careful grinding. Next, dop the stone. When it has hardened, and keeping within the outline drawn on the stone, follow the instructions for making a pebble cabochon, concentrating on gently and meticulously grinding the bevel surfaces away, so that they merge into a single, smooth curve.

A CABOCHON FROM GEM ROUGH

To make a cabochon from a piece of gem rough from the rock shop you will need to consider the stone carefully, in order to decide which will be the best way of getting a cabochon shape out of it, and which would be the most suitable area for the base.

The form will be conditioned by the state of the rough (it should not be chipped or fractured) and the area showing the most attractive internal pattern. If it is a piece of eye agate, for example, you might want to shape the dome so that the ring bands go round it.

When you have decided where the cabochon is to be sited, the first step is to grind the flat base, and then draw the chosen template shape onto it. The stone above should not slope off steeply, or you will end up with either a very lop-sided, or a shallow cabochon. Grind the sides to near the outline, and bevel the lower edge.

If the material is very jagged, you may want to grind away some of the rough angular pieces, but be careful not to whittle the stone away too much. Try to arrive at a basic, symmetrical shape.

Dop the stone, and proceed to bevel it as for the cabochon from a piece of flat stone.

A NOTE ABOUT FACETING

The other ornamental way of grinding and polishing gems is faceting. This is the method of cutting a pattern of flat surfaces over the stone, to reflect the light to the utmost effect, and enhance its beauty and brilliancy. Diamonds and sapphires, topaz and garnets are all treated with this process. Lapidaries have evolved cuts which are especially suited to the original crystal form—such as the 'brilliant' cut for diamonds, and the 'emerald' cut, a rectangular, stepped cut, for emeralds.

Faceting was an art practised only by the lapidary or very few deeply-interested amateurs in the old days. But today, with precision, do-it-yourself faceting equipment, amateurs are able to facet their own gems. Even so, they need to be amateurs with an out-of-the-ordinary grasp of the principles of the refraction of light and the technique of grinding and polishing precious and often small amounts of gem material.

Several books give advice and precise instructions to amateur lapidaries about faceting. A short list is given at the end of this book.

CHAPTER SEVEN

Tumbler Polishers

As you become more involved in the pebble hobby, and have perhaps polished one or two stones by hand, you will probably begin to wish that you could find a speedier alternative to this laborious method. The answer to your wish is readily available. The invention of the easily-operated, relatively inexpensive tumbler polisher has been largely responsible for the widely stimulated interest in the pebble hobby.

THE TUMBLER POLISHER

A tumbler polisher is a machine which shakes or tumbles a load of stones in a watertight barrel, the process of grinding them smooth and giving them a high polish being speeded up by the addition of grits and polishing agents to the water. It is, essentially the same process whereby nature smooths the rough fragments of rock which break off larger masses and makes curved and rounded water worn pebbles out of them, tumbling them in swift rivers and the breaking waves until their rough edges disappear in the slow passing of time.

Human beings, the impatient species, can't wait about like this, even if life were long enough, and so have ingeniously evolved the methods which have reduced the whole operation to about three or four weeks, from start to finish.

Properly carried out, at the end of this time, stones that went into the tumbler barrel looking weathered and discoloured, even hammer-broken, rough fragments of hard gem material, come out looking like glossy jewels or rare marble, including pebbles which we might kick aside indifferently as having no particular visual distinction.

If the stones were ground and shaped before going into the barrel, they are called 'preforms', and the irregular polished stones are called 'baroque stones', or 'baroque gems', if they are of gem-quality material. There are many ways of using these polished stones in jewellery and ornaments, and fittings and accessories to use for this purpose are sold in all the rock shops. They include ring blanks, cuff links, bracelets and ornamental brooch mounts, as well as fittings of other kinds, penholders and the like.

A glass kitchen jar with a lid, full of polished pebbles, makes an excellent present. So do ornamental arrangements of pebbles under glass similar to those decorative displays of mounted butterflies that the Victorians used to make. Ideas for things to make using the pebbles you polish in your tumbler polisher are given later.

Of all the equipment used in the pebble hobby, this is probably the machine which gives the widest pleasure.

TUMBLING MACHINES: A GUIDE

The most important point about the tumbler polisher you choose is its *efficiency*. You want a machine which will operate steadily, day and night, for weeks on end without breaking down, and you don't have to go to great expense to get this. Some of the most efficient machines, producing excellent results, which are on the market are the uncomplicated, simply-constructed ones and among the least expensive.

It is helpful to have refinements like variable-speed controls, and well-designed barrels which make the job of loading and releasing gas pressure easier, but first and foremost you need a machine which has been well thought out by its designers—one which has a good ratio between the speed of the motor and the tumbling operation of the correctly-loaded barrel.

Word-of-mouth recommendation is likely to be your best guide here, but if you don't know anyone with a tumbler machine who can tell you about his, go to a reputable firm which will be active in putting right any faults which may develop.

Size

As with all the other machines that are used in the pebble hobby, tumbler polishers vary in size from tiny versions which capably deal with about one and half pounds at a time, to massive industrial versions, which handle large loads of stones for commercial purposes.

For tumbling your own pebbles and gem rough at home, a machine which deals with about three pounds of stones in one batch will probably supply you with as many polished baroques as you can use. Before polishing, this would be about the amount that would almost fill a pint measuring jug. When polished their bulk will be reduced, to fill about four fifths of the jug.

Bear in mind that you will be repeating the process of tumbling, and that a lot of individual stones go to make up this amount, especially when you remember that you will want to be selective about the stones you polish from among those you find, and that buying gem rough can be fairly expensive.

If you think you are likely to use more than this quantity, it is a sensible idea to consider tumbling machines with twin 3 lb. barrels. Then you can polish six pounds of stones in one batch if you wish, or you can have one barrel loaded with stones at the grinding stage, and the second loaded with stones at the polishing stage, operated together from a single motor.

Whichever type of tumbler you choose finally, you will be wise to buy a second barrel to be kept exclusively for the polishing stage. It is nearly impossible to remove every trace of carborundum grit from the barrel after grinding, and it is vital that not a particle should get into the polishing load.

Noise

When you are choosing your tumbling machine, pay particular attention to the noise that it makes. If you can possibly hear a

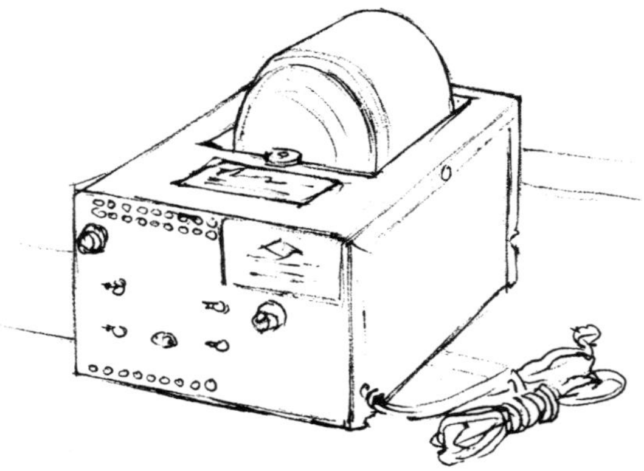

FIG. 27 A tumbler polisher

similar model operate, this would be best. But looking into the construction of the barrel will help you to assess this to some extent.

Not many small machines are on the market with metal barrels. These should have a rubber liner to reduce noise, as well as helping to keep down the possible build-up of gas pressure. Noise will be reduced in plastic barrels with rubber linings, or heavy-gauge, reinforced-plastic barrels. Using glass jars for tumbling barrels is not to be recommended. They are likely to explode if the gas pressure builds up inside them during tumbling operations.

Some manufacturers reduce noise by insulating the housing of the machine with a rubber lining. Look at the overall construction and method of operation of the machine to see what efforts the maker has put into reducing the noise output.

A machine in which the stones are being flung about fast at too high an angle, apart from the probability of damage to the fragile stones, may tend to be noisier than average. Nor is it necessary for the machine to have a noisy motor. It is quite possible to obtain models in which the motor runs relatively quietly.

All other things taken into consideration, if you live in a small flat in a block, or in a terrace house with walls through which the noise carries, it isn't going to be easy to site the machine out of earshot. It may be worth your while to pay more for the quieter machine.

Variable speeds
Some machines are fitted with variable-speed controls. The faster rates are for the grinding process, and slower for polishing. The speed at which the load is tumbled is of great importance, because if the motion is too fast, instead of travelling up the sides to the top of the barrel and tumbling down in a cascade, the pebbles may be flung about and crack against each other.

Too-slow speeds leave the pebbles slithering about at the bottom of the barrel, and the result is to produce flattened pebbles. In a fixed-speed machine which is operating correctly, it is possible to slow down the operation by adding fillers to the tumbling mix, especially at the polishing stage. These fillers, wood chippings, cork fragments, or crushed nut shells, act as buffers to ease the rate and sharpness of contact between the stones when a gentler motion is required. You can make them at home, but if you want to save yourself the trouble, rock shops supply commercial fillers of suitable texture and material for the purpose, which, for the amounts used in hobby tumbling, is a fairly inexpensive indulgence.

If, on your machine, you notice that pebbles such as quartz tend to produce internal cracks, or you see surface crazing on specimens, and you can't slow the motor speed, you could add a filler at an earlier stage. Even with a machine with variable-speed controls, you need to learn how your machine operates to avoid the faults of too fast or too slow, or too rough or too sluggish motion for the stones.

Barrel construction
Important general points to look at in barrel construction are the durability, and design for easy cleaning. The barrel will have to stand up to continuous pounding from some of the hardest material in nature, to begin with jagged-shaped, day and night, for weeks at a time. Thin plastic would soon wear through, unless the plastic is reinforced and toughened, and a hard-wearing type.

A solid rubber barrel, or heavy-duty barrel with a strong rubber replaceable lining and easy to operate lid have obvious advantages. It is tempting to think of buying a cheaper machine which works well and fitting superior barrels onto it, other than those supplied with it. This could work, but the chances are that the better quality barrel will only fit its own base machine, or that it might be too heavy for the smaller motor to cope with.

Easy cleaning is an important feature of a tumbler barrel. There are few things more clinging and penetrating than silicon carbide grit, and every trace has to be washed away between operations; coarse grit would contaminate finer grit operations, and it is so essential to keep any grinding grit particles out of the polishing operation, that most people keep a second barrel exclusively for this use.

A barrel with a cylinder shape and a close-fitting lid which slips over each end is one type of circular barrel on the market. This is easy to clean, and hard-wearing when made of toughened plastic, but the lids are somewhat tricky to fit until you get the knack, owing to the creeping nature of the grit, which tends to get between the lid and the cylinder, and the necessity of removing all air from the barrel, so that water isn't squeezed out. With this type of barrel, the trick of seating the lid is to clean up all surplus grit round the edge of the barrel, set the lid true on the cylinder, press it into position fully, then put both thumbs in the centre, pressing down hard. This expels any trapped air.

Types of tumbler

A popular design which is widely used for small- and large-sized tumblers, is a machine which has a barrel or barrels resting horizontally on spindles, which are rotated when the pulley attached to the motor in the base is set in motion. The correctly loaded barrel is pushed over slowly, and the pebbles are taken up with it, to fall back at the top in a cascade, rather like breaking waves.

Grinding and polishing a batch of stones with this type of machine usually lasts about three to four weeks, day and night. There is likely to be some loss of bulk in the stones.

Hexagonal Barrel Tumbler

This machine looks like a miniature concrete mixer. The barrel

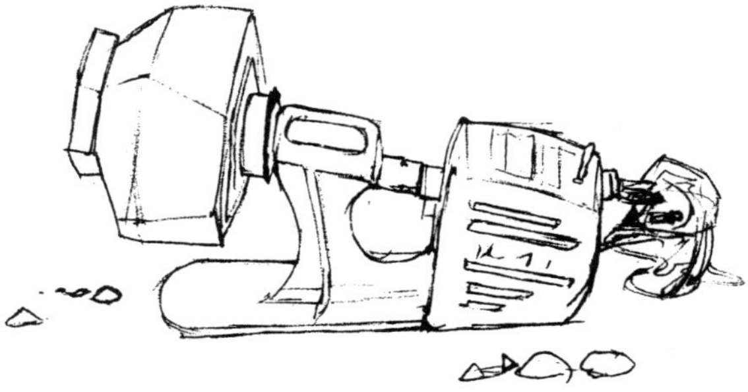

FIG. 28 A hexagonal barrel tumbler polisher

projects forward above the motor, and is hexagonal, with an open end. This prevents the build-up of gas pressure, which can happen in a sealed barrel, and also gives the opportunity of inspecting the stones frequently during tumbling operations without having to undo lids, etc. The angular construction is claimed by the manufacturers of some makes to reduce tumbling time.

Vibrating tumblers

These machines operate on an entirely different principle of tumbling. They don't, in fact, tumble the stones, they vibrate

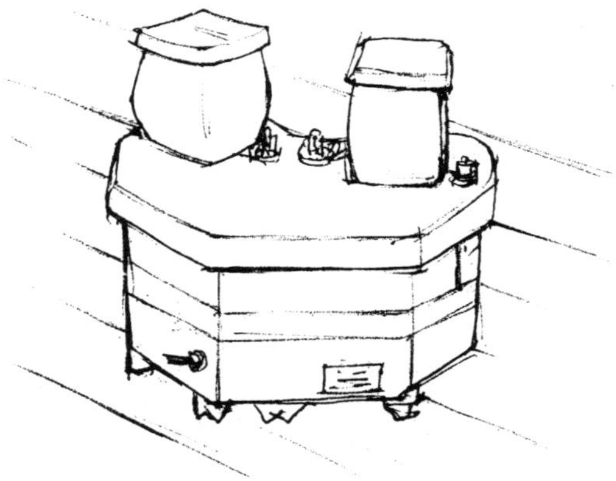

FIG. 29 A vibrating tumbler machine

them. The barrel is an angular hopper which stands on the base from which it is operated. The load of stones is shaken at high speed, with the advantage of quiet, almost motionless operations, giving very little loss of bulk and considerable reduction in the time taken to complete the tumbling process. Vibrating tumblers do not, however, come in the cheaper ranges of tumbler polishers at the time of writing.

PEBBLES FOR TUMBLING

If you have decided to have a trial run in your new tumbler using bought gem rough, some of the doubt about whether the stones will be suitable or not for polishing will be removed. The shop which supplies your tumbling grits will be able to supply you with the stones.

Choose a mixture of types and sizes, with as much variety as possible, to help you to become familiar with the different varieties. It is exciting to polish your own gems for the first time, in any case, and you should be able to have minerals like tigereye quartz, pale green amazonite, bright green aventurine, amethyst and agate, which will all tumble-polish readily.

When polishing your own pebbles, you have to decide first which are hard enough to stand up to the tumbling process, and which would crumble away to nothing. One man, enthusiastic about this new tumbler toy, took a knob off his marble mantlepiece and put it in the tumbler to polish, and that was the last he ever saw of it!

Spread the pebbles you have collected out on a newspaper on the table, to select a batch for polishing. Pick out those you would like to see polished and set them aside. Check them for hardness and, until you know your machine well enough to play around with it and see if you can polish softer specimens in shorter polishing times, keep to material over about 5·50 hardness. Leave out deeply-chipped or cracked specimens.

Hematite has been mentioned before for the red mess it makes when polished. Don't include any in a mixed tumbler load, or it may stain the porous stones and get into the crevices of others. Hematite is purplish-black-looking in fragments, and makes a red streak when you draw it across the back of a bathroom tile. Also, leave out pebbles which flake, crumble and have a low hardness number. These are usually sedimentary pebbles, and the more fragile crystal varieties. In choosing a load it is always better to keep pebbles of similar hardness, and the majority of similar size, together.

A good mixture will include the bulk of medium-sized pebbles, about 1 in. to $1\frac{1}{2}$ in. across, two or three larger pebbles (2–$2\frac{1}{2}$ in.), perhaps with an even larger pebble or two included, and an amount of small chips or large grit, tiny, round quartz pebbles, and chalcedony fragments found on the beach to act as a tumbling grit for the larger stones to grind and knock into their hollows and inside surfaces, and help carry the grit into them.

When you wash through your tumbler load at the end of tumbling, take care of these little stones. They polish well, and are useful for sticking on small presents (e.g., a jewelled matchbox).

THE TUMBLING PROCESS

To prepare for the tumbling process the first thing to be done is to set up the tumbler. If you have a suitable place to site it out of earshot, this would be better, preferably on the floor on a fire-resistant tray, in the garage or the junk room. If this is not possible, it could be put in a bedroom by day with the door closed, and the living room by night.

Check that all electrical leads, plugs and sockets are correctly connected and earthed, and that the motor is insulated from possible water penetration should the barrel lid possibly work off and spill its contents (this applies especially to home-built machines).

To use in tumbling you will need silicon carbide grit in 60, 80, or 120 mesh; 220 mesh for an optional second grind. However, this stage may be omitted, and you can go straight to 320 or 400 grit. Lastly, for polishing, you will need ceric or tin oxide polishing powder and some filling additive, which could be concocted at home out of nut shells, cork fragments or wood chips, or be a commercial filler.

You could put aside some old spoons for measuring the polishing and grinding agents into the barrel, and have a colander to place the stones in to rinse them through (into a plastic pail). The colander could be plastic or enamel, and should be kept exclusively for this use.

Wash the stones if they are beach pebbles, and fill the barrel two-thirds to three-quarters full with a mixture as suggested in the previous section, 'Pebbles for Tumbling', including some medium, some small and two or three larger stones. Cover them with water, and add the coarse grit. For a three-pound-sized barrel add about three to four ounces of grit, or two heaped tablespoons.

Fit the lid on securely so that no air is bubbling out with water round the rim. Wipe round the barrel, and check that both lids (if there are two) are correctly in position, and place the barrel horizontally on the spindles. (If you are operating a different type of tumbler follow the maker's instructions.) Check there are no leaks anywhere, and switch on the motor.

A daily routine to establish from the beginning, is to open the lid of the barrel at least once to release any gas pressure that may build up. This is a side effect of the grinding and polishing processes and can be dangerous enough to blow the lid off the barrel, if not to explode the barrel itself.

When you are satisfied that everything is operating correctly you can leave the polisher at work. Make a note on the calendar which date the operation started, noting the figure of the grit mesh being used. It is extraordinary how difficult it is to remember as the days go by, and you will want to know about when is the time to change over to a new grit.

At the end of twenty-four hours, switch off the electrical current and ease off the lid for the gas-release routine already mentioned. There is no harm in looking at a stone or two, but they will be covered in grey soup by now, and there won't be much to see. Replace the lid as carefully as when you set up the machine the day before, check for leaks, switch on, and let the tumbler carry on for another twenty-four hours.

If you should have to go away and switch off the machine for a few hours, make a note of it on the calendar so that you can add some extra tine to this grinding stage to make up for it.

Smooth, water-worn pebbles won't need quite so long to grind, unless any of them has a weathered coating which is difficult to remove. Inspect them after four or five days. Take a few out of the liquid in which they are grinding, place them in the colander over

FIG. 30 Bucket and colander of pebbles

a pail, and take them to the sink to rinse them through. If they have lost their weathered surfaces completely, except perhaps for white patches on flint which are difficult to grind away, but still look a little rough, put them back and continue grinding for a day or two longer.

Gem rough will almost certainly need at least seven days and nights in the coarse grind, because it is a hard material, and because of the rough state that it was in to begin with.

The aim of the first coarse grind is to wear the surface of the pebble clean of any rough patches and discolouration which is concealing its true colours and patterns.

After seven days

Test a sample of two or three stones before throwing away the grit. Let them dry after rinsing, before you decide whether they are ready to go on to the medium grind. A glance at them through a magnifying glass to make sure that no roughness remains would not come amiss.

If they are ready, hold the colander over the pail, tip the stones into it, and rinse them under the tap. When they are absolutely free of any grit particles, take the pail with the sludge of grit and stone particles in it, and throw the mixture away in an unused corner of your garden. Otherwise it should be allowed to soak through thick newspaper as a filter (direct from the barrel) and thrown into the dustbin.

Wash out the barrel and the lids thoroughly, and dry them with disposable paper or a spare cloth. Put some warm water and a dash of washing-up liquid in the pail, and wash the stones in it.

Return them to the colander and rinse them thoroughly. Already they should be beginning to look excitingly new and colourful.

Second stage

Now you can go on to the medium grind. This could be using 220 grit for four or five days, and changing over to 320 or 400 grit for another four or five days, but as the grit breaks down into smaller sizes during the grinding operation there is not a lot to be gained by this. Use the same amount of grit as for the 80/120 grind. Whichever you choose, this phase is extremely important. It is now or never that you will need to prepare the stones for polishing to a degree of near-perfect smoothness, with all tiny marks and scratches erased. Inspect carefully through the magnifying glass before going on. As with the other ways of polishing stones, the polish will not cure or conceal any flaws in the grinding. You may need to allow seven to ten days for this stage.

Polishing

Cleaning everything after grinding in a tumbler is particularly important, as the grit will be loose, not bonded in a wheel or onto wet and dry paper. It creeps in everywhere. Therefore, before proceeding to polishing, every single spoon, barrel, receptacle and pail will have to be cleaned out thoroughly, and the silicon carbide barrel put away.

Repeat the operation of washing the stones in hand warm water with a dash of washing-up liquid, and rinse them again.

Take the second barrel, and the polishing agent, and place the clean pebbles (they should, by now, be looking very attractive) in the barrel, which they will probably only about half fill, owing to their loss of bulk in the grinding process. Cover them with water, and about half the amount of ceric oxide to the amount of grit used.

Add a couple of tablespoons of filler and possibly a small amount of Polycell wallpaper adhesive, which helps to make the mixture a thicker consistency.

Polish for at least seven days. At the end of that time, if after inspecting the pebbles without throwing away the mixture in the barrel, they look glossy and beautiful, repeat the operation of rinsing them through thoroughly, using the colander over a pail, then get rid of the polishing waste from the barrel.

Wash all the equipment thoroughly and put it away for next time, if you are not starting to polish a new batch of pebbles straight away.

Now you can move on to enjoy the pebbles, placing them in a dish to be seen, or making things with them.

CHAPTER EIGHT

Things to do with Pebbles

The best way is to use pebbles very simply. Their attraction is in their natural appearance, whether enhanced by being polished or not. They can be kept in glass jars, or lying in dishes, and if unpolished look well in a pottery bowl covered with water, which will intensify their colour, as when lying on a stream bed or under the waves. Pebbles and mineral crystals can be displayed by various methods; mounted on a fabricated cube of transparent perspex, or wood, or coiled in wire, as natural sculptures.

Indoors, they can be placed in miniature gardens, under glass or as small-scale rockeries, or put here and there in the aquarium; striking, banded specimens look good like this. Out of doors, pebbles can be concreted in walls, in solid faces of cobbles, or in patterns, or they can be set in concrete in patios, as well as used loose.

For your own use, or as presents for others, the obvious things to make are necklaces, pendants and bracelets, using polished gem baroques or cabochons in settings. There is an enormous range of fittings for making up jewellery, called jewellery findings, including some of finer quality than the more mass-produced kinds. These can usually be obtained from a craft or rock shop, and are often illustrated in the firm's mail order catalogues.

All you need in addition for many of these fittings, is some of that remarkable tough contemporary adhesive, epoxy resin, which is a two-part mixture of resin and hardener. Araldite and Bostik make this type of adhesive. A couple of pairs of jeweller's pliers, and cutting and ordinary pliers are also needed. If you have the opportunity to use an enamelling kiln, this will enable you to make even more striking jewellery, with brilliant glossy coloured designs in lacquer, set with gleaming stones, preferably polished cabochon gems.

Soon you may want to go to jewellery-making classes to learn how to make up your own settings, using silver and other distinguished metals, but if you want to start making things immediately, the ready-made mounts, ring and brooch blanks, cufflink and earring settings and other items available require no special skill to assemble.

As a gift, a polished large pebble needs no explanation, and is sure to be received with pleasure. They also come in useful for paper weights.

Save all chips and tiny oddments of polished gem and flint that pass through the tumbler polishing process. Don't throw away the smallest fragment, but keep all of them in a polythene bag, adding any more from each polishing batch. Gem particles can be used for scattering thickly on a base of epoxy resin or tile cement, on box lids or jars, or carefully selected in colours and size and built up into a mosaic picture or design. The slimmest transparent sliver of agate has its use in jewellery as an insect wing or flower petal.

Arranging pebbles like the nineteenth-century decorative displays of butterflies, in a deep frame under glass, will give you opportunities to make use of common shingle, which is the bulk of the material that you will find. Suddenly you will discover that it is a positive advantage to have a number of stones which are similar in size and colour, which will enable you to make rows or bands of stones to form a pattern, brown flint next to iron-red-coloured flint; gold flint by white quartz. If you have enough of a selection of polished stones, ranging from about half-an-inch to an inch across, irregular or even-shaped, you can lay them out and cement them in star and snowflake designs, even in whirls graded to look like waves curling over, or in op-art spirals. A decoration using pebbles like this can be framed under glass and fixed to the wall, or cemented under plate glass on a table top.

There are absurd, delightful things that you can do with tiny pebbles, such as hanging them from the branches of a handsome twig to make a pebble tree, planted in plaster of Paris or cement in a pot. You can paint pictures on them, using oil paint or acrylic colours, and varnishing them. There is a tracing of a sailing ship in this section you can practice with, to discover how easy it is to do.

Pictures can be made as mosaics of stones, or from chips and hammer-broken fragments, or thin slabs of stone, depending not so much on the rarity of the rock material, as on contrasting colours and patterns in the stone, and its possibilities for polishing. They could be free arrangements of chips to form marguerites or amethyst violets, or carefully inlaid squares and shapes in stone, forming a flat surface and perhaps a picture—a process called 'intarsia'.

Apart from the adhesives and pliers needed, possession of a diamond saw for trimming and slabbing the flat pieces of rock, and possibly a diamond drill, with a diamond-impregnated bit and other accessories for enabling you to drill holes in stones, will greatly extend the things you can do with your pebbles and gems. It is all a matter of ingenuity and ideas; the scope is enormous.

A few suggestions for things to make from pebbles and rocks are given with simple instructions on the following pages, as well as information about tools and diamond drills and their use.

TOOLS, ADHESIVES AND SUNDRIES

A sliding gauge for measuring slabs will be useful if your saw is not fitted with a device for measuring parallel slabs. The *aluminium scribing pencil* used for marking cabochon shapes on stones is also needed here for marking positions of drilling holes and grooves for jump rings.

For painting pictures and patterns on pebbles, which should be unpolished and non-absorbent for the purpose, you can buy some *oil paints* and *picture varnish*, or, easier for the beginner to use, *acrylic paints* and *varnish*, with a *medium* and *fine brush*.

To mount items, mineral clusters or fine pebbles, *Perspex* or *wood*, *metal wire* or indeed, *stone* itself, can be used for a base. To bore holes through gem baroques, a *diamond-tipped bit* used in a *high-speed hand drill* is the most effective method available at a cost within the means of the average hobbyist.

Another pleasant way of working with pebbles, is to set them inside *clear liquid plastic*. 'Plasticraft' is the name of such a liquid plastic, and it can be obtained in kits or individual parts, with moulds, and colours which can be added. *Enamelling* mounts using a *small home enamelling kiln* is another variation, but bear in mind that stones should only be fixed to the mount after the enamelling process has been completed, or you may have the disappointment of seeing the mauve colour fade out of an amethyst, or spoil fragile gems affected by heat.

With an accumulation of a number of pebbles and rocks, polished and unpolished, of all shapes and sizes surplus to what you want to keep for your collection, you will doubtless be thinking of making some of them into jewellery or other items. Things which will, if possible add to the interest and pleasure of the stones, and set off their natural simplicity and beauty, even if amusingly.

If you use standard fittings and mounts for the jewellery you make up, the only additional tools you will need for the designs given here are two pairs of *pliers*, the slim-tipped type called *jeweller's pliers, one pair snipe-nosed, one pair flat-nosed*. Some *wire cutters* will also be useful to have by, and *tweezers* for lifting a small stone carefully and setting it in position in adhesive.

Adhesive is essential. *Two-part epoxy resin*, which consists of a resin and hardener mixed together carefully following the maker's instructions, produces a strong, powerful adhesive. Araldite is a well known and well tried epoxy resin. Another is Bostik epoxy adhesive; adhesives such as Bostik No. 1 *clear adhesive* can also be used for many of the sticking jobs when making up pebbles. *Tile cement*, such as Polyfix ceramic tile cement, can be used for bonding pebbles to larger, sound surfaces, and rock slabs to table tops, among other things. *A silicon carbide stick*, or *wet and dry silicon carbide paper,* to roughen smooth surfaces and aid adhesion.

Your *geological or crack hammer* and *cold chisel* will be needed for breaking up rock into chips for stone chip pictures, and your *diamond saw blade*, if you have one or the opportunity to use one, is required for cutting slabs for inlay work, and also for cutting grooves in the tips of pebbles for inserting and cementing in jump rings.

THE DIAMOND DRILL

Sometimes the only way to fasten a stone to a bracelet or the collar band of a choker necklace, is to bore a small hole through it, through which a link can be threaded. These links are little pieces of gilt or silver-coloured wire twisted into rings, called jump rings, which are opened with pliers to enable you to pass them through the drilled hole in the stone.

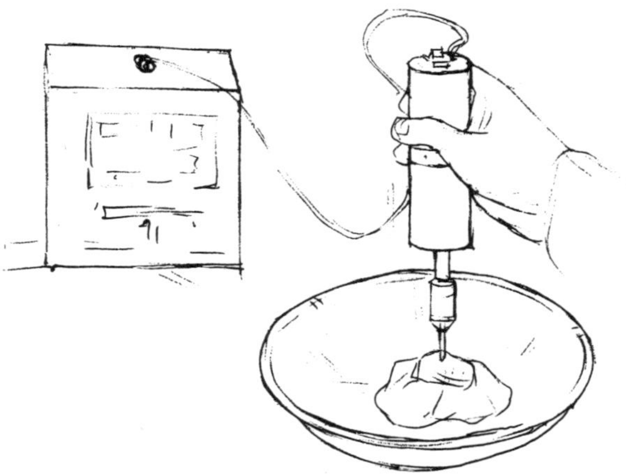

FIG. 31 Using a diamond drill

As already mentioned, the most efficient tool for drilling holes through small pieces of delicate mineral or rock material is a purpose-built, battery-operated, high-speed drill, fitted with a diamond bit. The Expo Titan high-speed drill, recommended here, is operated from a 12-volt or two 6-volt dry batteries, or from a car battery. At the time of going to press it can be obtained from some rock and lapidary machinery suppliers, such as Craftorama, 14 Endell Street, London W.C.2, and from Kernocraft Rocks and Gems Ltd., 44 Lemon Street, Truro, Cornwall.

The drill is easy to use and, unless it is operated in conjunction with a platform designed especially to hold the pebble in position under it, it is held in the hand. A diamond bit is inserted in the collet and both are then screwed into the drill. An important point to check before you commence drilling a stone is that the leads are connected to the battery the right way round. When using the drill, if you find that the screw holding the bit in position in the collet tends to work itself loose, this means that you have attached the drill's two electric leads to the battery the wrong way round. To reverse the direction of the spin of the drill, change the leads over, and clip or plug them into the opposite connections in the battery. This will make the screw tighten rather than loosen as it spins.

It is certainly wise to establish that this point is correctly adjusted before you begin to drill, or you could quickly ruin the expensive diamond bit if it begins to wobble or fall out while it is at work.

Diamond sintered bits, which are more extensively impregnated with diamond grit, are now available, as well as diamond-grit tipped bits. Sintered bits are more expensive, but have a longer life, if properly used. Diamond bits come in other sizes, and there are other tools which can be used with the drill for cleaning, polishing, brushing and burring.

If you do not own the platform for holding the stone in position under the drill, you will need a dish, some plasticine and some water. Embed the stone firmly in about half-inch deep plasticine, and stand it in the plasticine in the dish, covering it with a shallow layer of water. Hold the drill directly above it in a vertical position, and maintain this during drilling, avoiding allowing it to wobble. Follow the maker's instructions exactly as to times given to drilling and resting the drill.

For marking stones, use your aluminium scribing pencil.

MAKING A PEBBLE CHARM BRACELET

This charm bracelet illustrated (fig. 32) was made from miniature pebbles brought back from a trip to Crete as souvenirs. They include a tiny pebble of heliotrope, which is dark green, red-blotched chalcedony, known as bloodstone; some iron-stained quartz, and a small piece of conglomerate. Every pebble is different, and none is larger than three-quarters of an inch long by three-eighths of an inch across.

The pebbles are hung along a plain gilt chain which has fairly large links, by means of jump rings cemented into grooves cut into the top of each stone with the diamond saw. (If no diamond saw is available to enable you to cut grooves in the stones, you could substitute bell caps for the jump rings.)

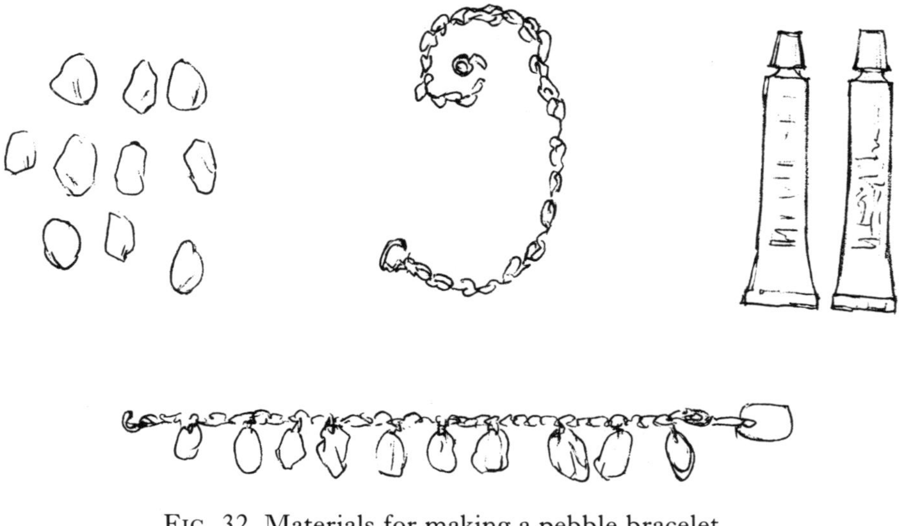

FIG. 32 Materials for making a pebble bracelet

Materials: a collection of small stones less than three-quarters of an inch long. About eight is a suitable number for a bracelet, unless you like them very crowded, when you will need more. The pebbles can be polished or unpolished. A jump ring for each pebble, and a short length of chain to fit round the owner's wrist. To measure the amount needed, take a tape measure closely round the wrist, and add about an inch, possibly a little more if wished. This will give you the length of chain required. (Don't make the bracelet too tight, or too uncomfortably loose.) Also, a hook to fasten the ends of the chain together when worn.

Tools: two pairs of jeweller's pliers; aluminium scribing pencil; two-part epoxy resin adhesive; a diamond saw; and plaster of Paris.

To make the bracelet:

1 Mark each pebble at the centre top, using the aluminium scribing pencil.
2 Embed each pebble in a small individual block of plaster of Paris, so the aluminium mark shows. Leave to harden. When fully set, clamp the first pebble in its cast into the rock clamp.
3 Saw a small groove in the top of the pebble, to a depth between one eighth and one quarter of an inch. Repeat for all the pebbles.
4 Carefully lever and chip off the plaster of Paris cast. Wash and dry the pebbles. Atrange them in their chosen order.
5 Using the pliers, open the jump rings, by twisting the wire

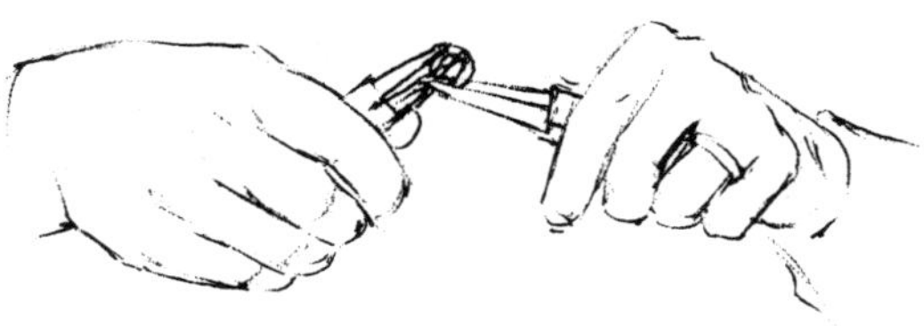

FIG. 33 Opening a jump ring with two pairs of pliers

apart at the join, the left hand pliers pulling the wire forward, and the right hand pliers turning the wire back.
6 Thread the jump rings through the chain, at regular intervals for a charm bracelet with the stones spaced out. For example, you could fix a jump ring through every sixth link of the chain, which could amount to them being about one inch apart. Close the jump rings by twisting the ends back together again, using the pliers.
7 Mix some epoxy resin and hardener as directed by the makers.

Using a small tool such as a screwdriver or nail, press some of the mixture into the groove of the first pebble.

8 With the chain lying flat on a tray, press the pebble by its adhesived groove onto the first jump ring.

9 Take the next pebble in order, and repeat the process. Continue until all the pebbles are fixed to the jump rings.

10 Without dislodging the pebbles, arrange them carefully on the tray, and place them where they may stay for three days, or the length of time required by the maker for the adhesive to harden. Leave them to set.

11 When the bracelet is securely hardened, open the hook using the pliers, and fasten it to one end of the chain.

SOME OTHER JEWELLERY

Some jewellery can be made by fastening polished stones to ready-made mounts by means of epoxy resin adhesive; others use jewellery findings fitted with prongs which are twisted over the stone. Jump rings passed through holes drilled through the stone are an alternative to setting a jump ring with adhesive in a groove

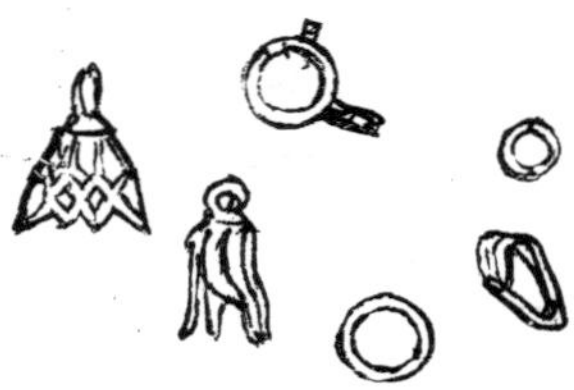

FIG. 34 Jump rings

in the baroque. It is even possible to buy a cage fitting to fit over the stone.

Ideas for using stones in these ways are given here. They are only a few of the things you can do with them, and are among the simplest methods.

Gem stone Pendant

A pendant can be made using a polished, ornamental gem stone—for example, a phantom amethyst polished baroque. To make it, you need an attractive flat stone picked out of a tumbler-polished batch; a 24 in. neck chain with a bolt ring clasp; a large, round jump ring, with a second in a smaller size; pliers and a drill with a diamond bit. Plasticine, water and a dish, if you do not own a platform to use in conjunction with your drill.

To make the pendant, first embed the stone in half an inch of plasticine, and place it in the dish, covered with water. Proceed to drill a hole through the stone below the centre of the top, just far

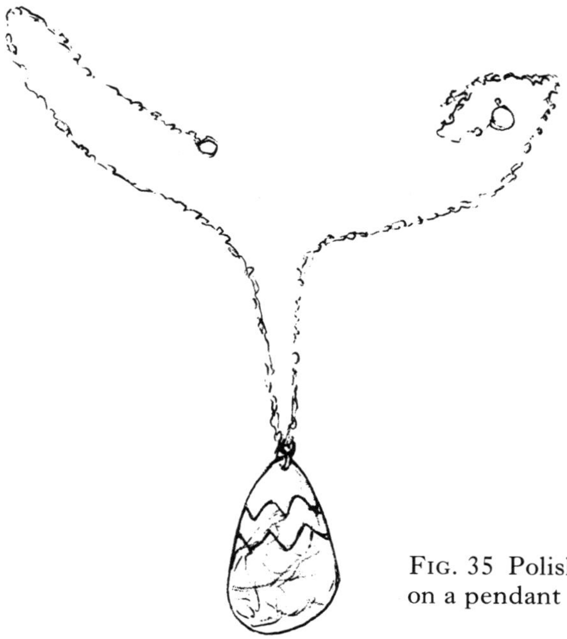

FIG. 35 Polished amethyst on a pendant

enough from the edge to avoid breaking the stone. Open the large jump ring with pliers, in the way described on page 128, and pass it through the hole in the gem, and close it again.

Take the second jump ring, open it using the pliers, and pass it through the first jump ring, and over the chain. Close the ring ends together. Tie the chain in a knot, or wear it loose.

Pebble Collar

A collar of polished pebbles hanging from a leather choker band is another attractive thing to make. Use a strip of leather to fit your neck, not too tightly, with small holes punched through it at equal intervals across the lower edge and at the ends. Drill holes through the centre of 6 to 8 small, polished stones, fairly close to the top. Attach the stones by jump rings to short lengths of chain detached from the main length by pliers, or several jump rings linked together, and fastened to the collar strip. Tie a leather thong through the holes at the ends of the band as a fastening.

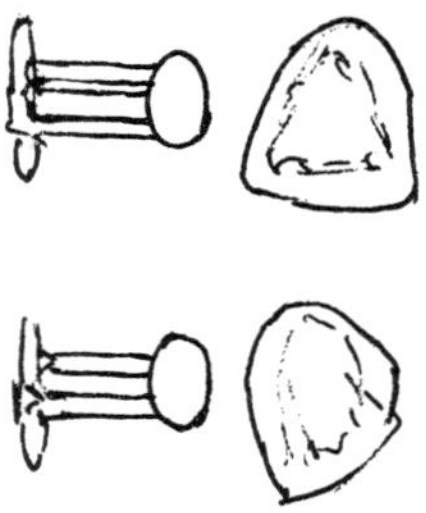

FIG. 36 Cuff-link mounts and pebbles

PEBBLE NOTIONS AND PEBBLE PRESENTS

Glass jars, bought from china and kitchen equipment departments of stores and hardware shops, can be filled with polished pebbles and given as presents, or kept at home as ornaments.

Old stoneware storage jars or flagons or marmalade pots, earthenware dishes or jars, can also be decorated with pebbles stuck on the outside, arranged in patterns, bands or spirals, or thickly cemented on. All you need are the jars, bowls or pots, some epoxy resin or tile cement, small, well-rounded, polished flint and quartz pebbles, a felt-tip pen, a ruler and a tape measure.

Use the felt-tip pen to make the top edge of the jar in four equal parts. Put a mark at the lower edge of the jar exactly under each of the marks on the rim. Join the marks diagonally with dots across the side of the jar.

Choose the most decorative side of the pebbles for the upper side; mix the resin and hardener of the epoxy resin as directed by the maker, and when prepared, put a dab on the wrong side of each pebble. You will probably need about thirty for a medium small jar.

Put dabs of epoxy resin along the lines of dots on the jar. Leave the pebbles and jar to dry for the time specified for the adhesive

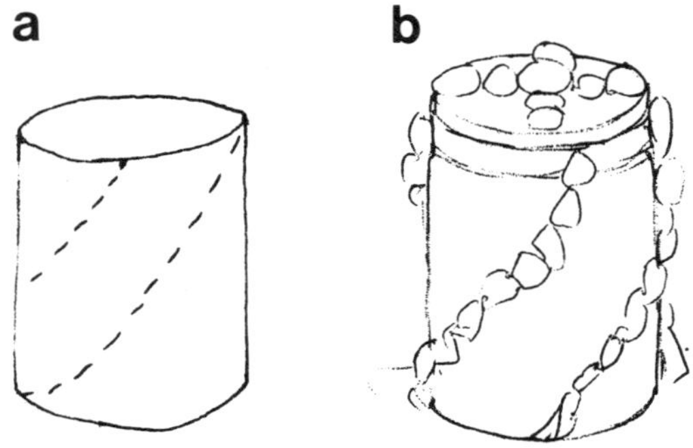

FIG. 37a Marking a jar for pebble decoration
(b) The jar ornamented with pebbles

used. When the time has passed, press the pebbles onto the adhesive firmly, selecting as you go the most decorative colour arrangement for them. Stand the jar aside for as long as the makers' suggest that the adhesive needs to harden, probably about three days.

Repeat the process to stick the pebbles to the lid, but instead of diagonal bands, make a cross of pebbles which quarters the lid from edge to edge. Choose round, fine-coloured and shaped pebbles for the lid, with perhaps a graded variation in sizes. The pebbles could be arranged in rings, or vertical lines too.

A shingle-covered lamp base

A jar or bottle which is a suitable shape and size for a lamp base can be covered with Polyfix ceramic tile cement, or a layer of prepared two-part epoxy resin, and set with a quantity of polished small pebbles or natural shingle.

If the jar is not stable enough, it can be made heavier by setting some plaster of Paris inside it, or half-filling it with some extra stones. A socket for the bulb which will fit into the neck of a bottle, with an electric lead from it, can be obtained from electrical equipment shops.

MOSAICS FROM GEM FRAGMENTS

Each batch of gem rough and beach pebbles polished in your tumbler is likely to produce a number of pretty fragments. Some

FIG. 38 Pebbles on a lamp base

are thin, almost-transparent polished slivers which were flaked off when the rock or mineral was crushed or hammer broken; others are irregular-shaped beans of gem stone; others, little dots of lapis lazuli, or flint or quartz.

Save them all. Indeed, as it helps the polishing process to include grit-sized stones with a load, to grind into the hollows of the larger pebbles, you can add some deliberately, and look out for them when you are picking up pebbles. When you wash through the newly-polished pebbles, collect these fragments, dry them, and add them to any more you have, kept in a polythene bag.

While you are about it, save any attractive little wooden or glass containers that come your way. Keep any wooden chocolate and cigar boxes too.

A Jewelled Matchbox Cover

Take a new box of matches; some mounting card; gold paper, lacquer, paint or P.V.C.; epoxy resin adhesive; a sharp knife, scissors, a pencil and ruler. A selection of equal-sized, though not necessarily equal-shaped gem or mineral polished stones. Pieces about $\frac{1}{4}$ in. across would be suitable.

Remove the machbox cover, and put the matches in the box to one side. Cut the matchbox cover open along one side and open out flat. Place this flattened cover on the mounting card, and mark out an identical sized piece of card, with lines for the angles where the

FIG. 39 Gem rough mosaic on match box

cover is folded. Add an extra amount as deep as one side of the matchbox cover, for overlapping when glueing the side. Cut out around the outline of this shape.

Using the sharp knife and ruler, cut carefully half way through each fold line drawn on the shape. Fold the card to form the same shape as the original matchbox cover, and glue the overlapping sides together, one on top of the other.

If you intend to paint or lacquer the cover, now is the time to do so. Give it an even coat, and set it aside to dry. If you have decided to use paper or P.V.C., cut out another shape to match the original cover, and paste adhesive over it on the wrong side. Paste adhesive over the card cover too, and stand it on one end until it is ready to be bonded with the paper or P.V.C. After the right amount of time, smooth them together.

Now you can select the polished mosaic to go on the top. It could consist of fine, varied examples of different kinds of stone, or it could be tiny fragments arranged in a picture or patterns, or even a skilful combination of large and small pieces in two or three colours only. The important thing is not to fling the stones on willy-nilly, but to have some idea and form in their arrangement.

When you have decided on the arrangement, lay the mosaic face-down on one side, and cover the reverse side of each stone with adhesive. Cover the top of the box with adhesive also, and leave both the stones and cover to dry for the specified time.

When they are ready, carefully and neatly press the stones in position on the lid, and leave the whole to harden for a few days. Lastly, cut out the surfaces for striking the matches from the original matchbox cover, and stick them on the sides of the jewelled cover. Put the matches and box inside the new cover.

Other Uses for Mosaics

Mosaics can be used in many ways. In jewellery, for example, a ready-made brooch mount can be covered with small, polished stones in the form of a butterfly. The body and head are made of blue lapis lazuli and green aventurine; the wings are rock crystal and rose quartz edged with amethyst. The spaces are filled in with minute pieces of darker stone, to set the lighter colours in relief.

Another delightful way of arranging gem fragments is to treat them three-dimensionally. Thus, a plain lipstick case can be set with a mosaic of flat stones on top, and a petal-like arrangement

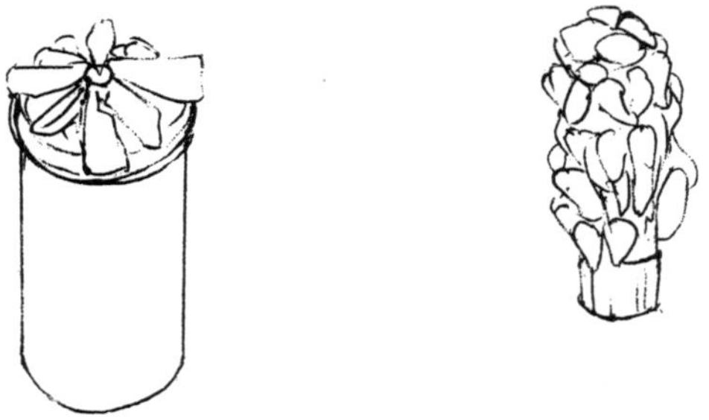

FIG. 40 Pill box and lipstick case with mosaics of gem fragments

down the sides. A row of flat stones around the top is overlapped by longer, transparent stones, which in turn, can be leafed by shorter flat stones around the lower edge of the case.

Powder jars and trinket boxes can have three-dimensional designs on them, and tiny frames can be set with pictures that stand out. A jewelled insect makes a fascinating decoration; its wings can be made of translucent agate and chalcedony.

PAPER WEIGHTS AND POLISHED STONE

Large polished pebbles, beautiful in themselves, are likely to be used as paper weights. For the same purpose, or purely as ornaments, one or several pebbles can be set in blocks of clear plastic, made by pouring liquid plastic with the addition of the appropriate amount of the correct hardener into a mould. The technique for using liquid plastic is given with the maker's instructions.

The effect of small pebbles or gem baroques floating inside a transparent, glass-like form is uncanny, and attractive. Clock parts, shells and other items can also be used in this way, too. Ceramic moulds are obtainable for making various shapes, including paper weights, key ring discs and pendants, or you can use any suitable container of your own, such as a glass dish, for a mould. As you become more expert, you may even want to attempt to plant irregular-shaped polished rock pieces in a larger shallow area, to set in a table top.

Book Ends and Candle Holders

Simple, polished blocks of stone can be converted into book ends, by cementing or screwing a narrow-gauge sheet of soft metal or thin wood under the base, which will extend under the books. Blocks of polished rock like this can also be used as mounts for mineral clusters of crystals, or an exceptionally fine pebble.

A stone with a hole bored in the top, into which a small sharp nail is cemented, point uppermost, can be used as a candle holder.

PEBBLE DECORATIONS

If you own a glass-topped display cabinet, it is a simple matter to lay pebbles out in it. They can be displayed in different ways, fixed to a backing and framed, and hung on the wall. This can be a method of showing the different types of pebble, but is equally good for displaying the decorative qualities of polished stones.

All that is needed is a deep frame, a strong supporting backing, adhesive and a number of polished pebbles or gem baroques.

A polished pebble display

Materials needed: a frame with sides at least one inch deep, which could be made from wood, aluminium or steel edging, or plastic, painted or gilt frame-maker's framing. Stiff card or thin plywood for mounting the pebbles on, and thin felt or baize or loose sand, and adhesive. A ruler and pencil.

Stones to be mounted, which could consist of a number of polished flint and quartz pebbles, about an inch of less across, or gems to be arranged by colour, or variety.

Method: measure the frame, and cut out mounting card to fit. If this card is flexible, a second backing will be needed, to be screwed onto the frame, as extra support. Cut felt or baize to fit the mounting with a small margin to lap over the edges onto the back. Spread adhesive evenly all over the card and smooth the material over it. Alternatively, if sand is being used, cover the mount thickly and evenly with adhesive, and shake sand all over it, moving the card to and fro to spread the sand, shaking off the loose surplus.

Select the stones and decide on the arrangement. For a first attempt, simply mounting them in rows in order of colour will look effective.

If the frame is rectangular, either in a landscape or an upright shape, you could experiment with laying horizontal or vertical rows of pebbles an equal distance apart on the mount to see what they will look like when fixed. A row of pale grey pebbles could go beside a row of golden-speckled flints, which in turn could be next to white quartz, beside iron-stained flints, and so on. If the frame is round you could arrange the pebbles by colour in concentric circles, or grade them by colour, harmonising or contrasting them in a spiral pattern.

They could be laid out like a wheel, with spokes like a wagon wheel, and a fine circular pebble for the hub, and others for the rim. Or they could be in the form of a geometric star or snowflake crystal.

The arrangement decided on, a tracing will have to be made of the position of the stones. Carefully, they could be rearranged on tracing or transparent greaseproof paper, and sketched round, then laid to one side in order, retaining their formation. After this their layout as sketched on the paper will need to be checked with the ruler, to make sure that the rows are truly straight and an equal distance apart. A pair of compasses will help you ascertain that any ring is accurately circular.

Cut a scrap of paper out of the centre of each pebble tracing when their position has been adjusted on the layout and, using either chalk or a felt-tip pen, transfer the mark on to the prepared mount. Prepare the adhesive, and choose which side of each pebble will lie uppermost. Dab some adhesive on the reverse side. Place some adhesive on each spot marked for a pebble on the mount. Be generous, but do not let any spread between the spaces on the felt or sand, or it will show in the finished result.

After the adhesive has been left for the time required by the makers, press each pebble in position on the mount. Leave them to set and harden thoroughly for several days.

When the pebbles are firmly fixed to the mounting, set it in the frame under glass and, if necessary, add extra support across the back of the frame. Put an eyelet screw in each side of the frame, and a taut string across, using strong picture-hanging cord. Prepare a strong hook and hang the frame with the pebbles inside. The flatter you can fix the decoration to the wall, the better.

A display of gem baroques or cabochons would be superb in an elegant polished frame under glass; alternatively, use cut and polished specimens of banded or marked gem stones, arranged decoratively.

A STONE CHIP PICTURE

You can use your imagination when making a picture out of stone fragments. If you have a rough piece of red serpentine, and some green mottled serpentine, too, with possibly some white dolomite or pale feldspar rock, you can smash pieces off, to form chips and flat slivers, which you can freely arrange as leaves and flowers. You can include some polished fragments if you wish, to complete the effect you are aiming at.

Materials needed: a small frame, and card to fit it; epoxy resin adhesive; sand; a bottle of coloured acrylic ink, green or pale blue, or whichever colour you prefer; some dark green ink or fine cord; brush; pieces of soft, colourful rock, such as marble or serpentine, and a few polished gem rough pieces.

Tools: a hammer and cold chisel (a crack hammer is most suitable) and thick newspapers; goggles.

Method: cover the card backing with adhesive and shake sand over it evenly, scattering loose sand away; set aside to harden the adhesive thoroughly. When hardened, paint it all over with coloured ink.

Wear goggles, and cover the rock pieces with newspaper. (Lay them on earth, not a patio or floor, or you may crack more than the rock specimen.) Freely smash the rock pieces into smithereens, or if they cleave easily, find the point of cleavage with the cold chisel, and separate them into thin layers, which can be chipped into shape.

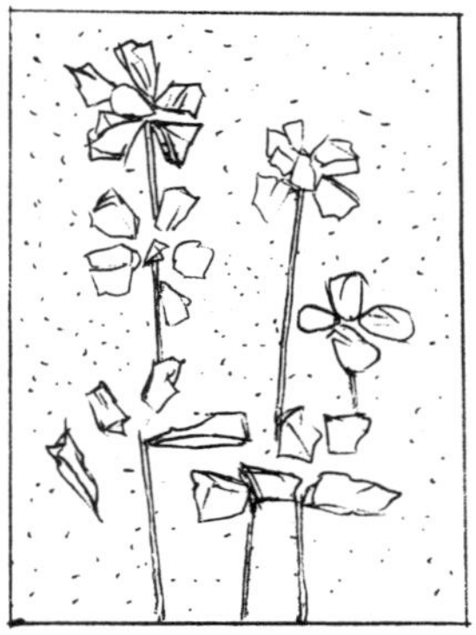

FIG. 41 A stone chip picture

Select flat, suitably shaped chips for flower petals and leaves, and a couple of small chunky pieces for the flower cntres. Arrange two or three flowers, with, say white petals, as marguerites, or small polished amethysts as violets, in a floral composition. Use green or dark rock fragments for leaves.

When the picture is arranged, trail some cord with adhesive on one side as stalks from the flowers to the leaves. Put adhesive on the back of each petal, and leaf, and the flower centres and some adhesive underneath on the mount. After the adhesive has dried for the specified time, press the flowers and leaves into position. Leave to harden for several days. Then fit the mount inside the frame, and hang it on the wall.

Using stone chips you could make pictures of birds on flowering branches, or wild flowers by a river, landscaping the scene with small pebbles where appropriate.

A PEBBLE TREE

If you can have rock-candy mountains and a lemonade tree, why not a pebble tree? It is an amusing bit of nonsense, but charming too.

Materials needed: a branched stumpy twig about eight inches tall, and seven to eight inches broad, stripped of leaves and berries.

FIG. 42 A pebble tree

Depending on the thickness of the twig branches, you need as many jump rings as you have pebbles, with possibly a second and third to hang them from the branches; a small flower pot about $3\frac{1}{2}$ in. to 4 in. in diameter, and plaster of Paris or cement; sand; adhesive; a number of small flat pebbles, or pear-shaped pieces of gem rough, about $\frac{3}{4}$ in. long. The pebbles can be polished or unpolished.

Tools: pliers; the diamond saw. (If no saw is available, you could buy slit stones, available from Ammonite Ltd., or use bell caps as an alternative to embedding a jump ring in each stone.)

Method: Embed each pebble in plaster of Paris, with the place to be notched showing. Clamp the pebble in the saw clamp, and cut a groove. Repeat for the other pebbles. Chip and lever off the plaster of Paris carefully. Next, prepare epoxy resin adhesive, and fill the groove in each pebble with it. Press a jump ring into each one. Leave to harden for the time specified by makers.

Meanwhile, fill the flower pot with plaster of Paris or cement, and plant the tree in it. Scatter a thick layer of sand on top of the plaster or cement round the foot of the 'tree', and leave all this to set and harden too. When both the tree and the pebbles are ready, pass a second jump ring through the first, and over the branch, using the pliers to open it. If necessary use a third jump ring to link the other two. Repeat for all the pebbles.

By boring holes through the top of flat polished gem fragments, especially those of rose quartz or rock crystal, and wiring them together, looping the wire round the branches, it is possible to add blossom to the pebble tree, or substitute it, and add leaves of thin polished gem material such as amazonite or aventurine.

PAINTING PEBBLES

You may or may not regard yourself as an artist, but painting a pattern or a simple picture on a pebble is not as difficult as you may think, especially if you help yourself by making a tracing from some other image, and cutting out your own stencil. You can easily paint patterns using cut out geometrical shapes, triangles, squares, circles, bands and wavy lines; dividing the pebble into coloured halves like the jester's costume, or harlequin diamonds.

As an introduction to painting a picture, you can make a stencil of a three-masted sailing ship. Trace it and cut out the shapes, and you can transfer the picture to your pebble.

Materials and tools needed: a large, smooth unpolished pebble; tubes of acrylic paint in white, and pale blue, a paint brush, the artist's type of brush, in a medium size, and a smaller brush for the fine lines. Varnish for the acrylic paint. Tracing paper such as greaseproof paper, and a felt-tip pen. An aluminium scribing pencil; scissors.

Method: trace the outline of the shapes of the sails and hull of the ship. Carefully cut out the main shapes. You can copy the masts and waves easily, freehand.

Place the stencil over the pebble, in a central position, so there is an equal space on either side of it and above and below Lightly

FIG. 43 Tracing for a ship painting on a pebble

mark the outlines of the ship onto the stone, using the aluminium scribing pencil. Lift off the stencil. Paint the sails and the hull in white; when they are dry, paint the masts in white, and the flag and portholes in pale blue. Copy the waves onto your stone in pale blue. When the picture is entirely dry and hard, varnish the stone, first on one side, and allowing it to dry, then on the other.

You can make stencils from other pictures and transfer them to smooth unpolished pebbles in this way. A simple picture of a bird or a plant or a scene of water lilies and bullrushes and a pond could look beautiful. Trace the picture lightly, not to damage it, and make notes about the colours by the tracing. Obtain acrylic, or if you feel you could cope with them, oil paints in the colours needed, and follow the same procedure as for the sailing ship.

Make a stencil by cutting out the main shapes, and transfer the picture to your stone using the aluminium scribing pencil. Paint in the large areas first, leave them to dry; follow with more superficial ares; end with the details. Varnish when dry.

Do not attempt to trace from a valuable picture or rare engraving.

A museum postcard often has just the picture to suit these requirements.

PEBBLE MOSAICS

The art of making mosaics, which is fitting together small pieces of stone, tile or glass to form a pattern or picture, has been known from early times in man's history. Mosaics have been discovered in the tombs and palaces of the Egyptian pharaohs; Roman mosaics are famous, pavements of small different-coloured pieces of marble, or in some cases in glass, which form magnificent pictures of gods and goddesses, people and animals, bordered by intricate patterns. Several such pavements still exist on Roman sites in Britain.

Mosaics have been made from that time to the present day. They have been laid over walls and floors, or set in jewellery, using minute stones no bigger than the hundreds and thousands used to decorate cakes, and matched into exquisite pictures or miniatures.

You can make your own mosaics, using small stones up to about an inch across, or larger squares and slabs of stone. The pieces should be about equal in size, though they can be irregularly shaped. This applies to pebbles too, and following the method for making a simple mosaic, you could create a patterned table top or

FIG. 44 A mosaic brooch

wall picture, or even make mosaics in some parts of your house on the wall or built into the fire surround.

You will need a piece of plywood or aluminium to fit the area you have in mind. Some people add an aluminium frame around the sheet, but this is optional. It gives a tidy edge to the work. The adhesive used can be epoxy resin, or Polyfix ceramic tile cement, but if you are using aluminium, or a metal base, it is better to use Instant Polyfix, an adhesive that dries clear, and does not crack with the expansion and contraction of the base.

From a large quantity of pebbles, select those you will need for your design. You may let this emerge from the assortment of pebbles you have to choose from, or you may have a sketch of a picture you would like to achieve if possible, with perhaps a few minor modifications. Whichever you have in mind, begin to lay the pebbles on the base, placing them for the best effect, gradually building up the picture or pattern as you arrange and rearrange them. This may be a slow process, but it is a rewarding and leisurely pursuit in an age of speeds faster than sound.

When you are content that you have achieved the arrangement you were aiming for, simply remove part of the picture and apply the Instant Polyfix or other adhesive a piece at a time, according to the maker's instructions. Usually you will have some time before the adhesive hardens to shift the mosaic pieces around a little.

When the mosaic is completed satisfactorily, leave it to harden for the length of time specified. If it is used in a table top, it will probably need a close-fitting, plate-glass cover in a well-supported frame.

Much the same method is used when forming tiny pictures of mosaic stone fragments, but you will need to handle the stones with tweezers.

If you have been to many different places collecting rock specimens you will probably have accumulated a number of pieces which have considerable sentimental value, apart from being varied in colour and texture. These can be slabbed and fitted as a rough mosaic on a low table, or the slabs can be ground smooth and polished, and fitted together in mosaic form.

Out of doors, pebble mosaics would need to be set in concrete reinforced with wire netting. This is a more specialised craft, but it can be carried out by an amateur with advice from a builder.

Another form of mosaic is carefully inlaid stone marquetry, called 'intarsia'. This is the name of an Italian method of inlaying flat, thin slabs of marble or other stone which are shaped so they fit together neatly, so the join can barely be seen.

The colours and internal patterns, the dark and light of the stones, are all taken into account when they are set in the picture.

FIG. 45 A stone inlay picture

Intarsia can be set in stone, or inlaid on wood or other backings. It is fairly difficult to do, because of the accuracy and care needed when sawing the parallel stone slabs, and in making the shapes fit together so neatly, so the join can barely be seen.

If you would like to attempt some intarsia work, begin with an ultra-simple design, and on a small scale. Intarsia is an art which can have extremely beautiful effects.

To make a simple intarsia picture

Materials needed: several pieces of different coloured marble or serpentine rock; a base of plywood or sheet aluminium, or possibly even some slate; epoxy resin, Instant Polyfix or other adhesive; and paper.

Tools: a sheet of plate glass; the diamond saw; an iron lap to fit over your lapidary machine, or a lap wheel. Grinding and polishing materials; a pencil; and an aluminium scribing pencil.

Method: draw the picture you would like to make, bearing in mind the choice of stone you have, and its colours when polished. Begin by preparing a number of thin, parallel slabs of stone, roughly about the amount to cover the area you estimate you will need for your design. Lay the pieces of slabbed stone on your drawing to assess the effect of their colours and patterns. Decide which parts of stone you will select, and where you will put them.

Trace the drawing, and stick the tracing to thin card, such as a cereal carton. Cut out the parts of the drawing, and then cut the parts into smaller shapes which fit over the slabs available, carefully positioning them by pattern. Lightly mark round the edges of the shapes, using the aluminium scribing pencil, on the slabs.

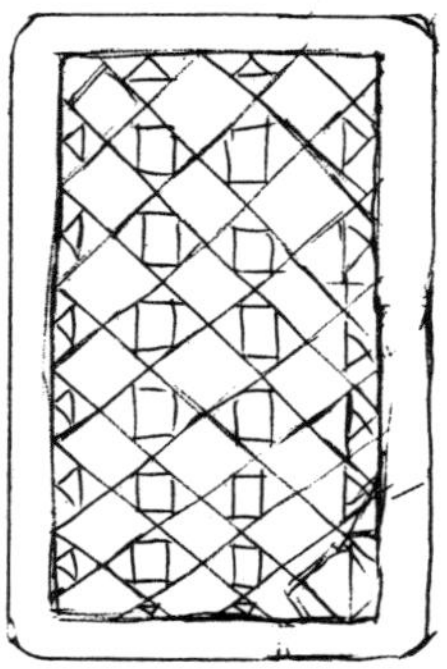

FIG. 46 Harlequin patterned stone inlay

Trim the slabs to these shapes, until all the pieces have been prepared. Lay them together to see how they fit, and replace any failures by trimming other pieces. Edges can be smoothed, using the attachments to your diamond drill, if you have one, or the grinding wheel, gently. When all the pieces fit together perfectly, the right way up, turn them face down on the plate glass.

Apply the adhesive as directed by the makers, and press the backing onto the intarsia. Turn the inlay up, make sure all joins are perfectly matched, and wipe off any surplus adhesive which may have seeped through. Leave to harden. When this process is complete, lap the picture and polish it.

A PEBBLE GARDEN

Pebbles can be used to decorate a miniature indoor rockery planted in a lead-lined or plastic container. Many different sizes and shapes of such containers are available today.

The container should be capable of holding potting compost about three to four inches deep, and, remembering that your aim includes displaying your pebbles among the plants, it is essential to choose plants which do not grow fast, or have luxuriant foliage which would conceal them.

Probably the most suitable plants to choose are slow-growing succulents and cacti. These are plants which store water in their leaves, but although cacti are succulents, not all succulents are cacti. They require a sandy soil, about half John Innes potting compost mixed with half coarse sand.

Plant the cacti and succulents first, then arrange the pebbles among them. In the summer they will need watering two or three times a week, enough to keep the soil moist, and should be placed in plenty of light, in a window. In the winter, they will only need watering about once every ten days or fortnight.

FIG. 47 Cactus indoor garden with pebbles

There is a wide range of cacti to choose from, such as small prickly pear, lobivia, and necporteria aspillaga. Among the succulents, suitable plants include crassulas and aloes. *Aloe variegata* has leaves shaped and patterned like partridge feathers. A rather interesting small plant that can be grown from seed, called lithops, is a little South African succulent that does in fact look like a pebble. Its name actually means 'pebble' too!

You could, if you wish, arrange a small garden under glass. The plants and pebbles could be used in a container covered by a glass top or dome, of the sort used by the Victorians to cover stuffed birds or wax flower arrangements. The plants will need headroom to grow, so bear in mind their eventual height when you plant them.

Bulbs can be grown in a dish of pebbles and water, and boldly-striped or banded pebbles, placed here and there on the floor of an aquarium, will add to the effect of a rock pool behind glass.

A garden centre will usually advice you on other plants to choose, and various books on house plants give information on gardens in containers, and the care and management of indoor gardens. A couple of useful books on the subject are included in the further reading list of books at the end of this volume.

PEBBLES IN AND OUT OF DOORS

Pebbles, and their larger counterparts, cobble stones, have always

FIG. 48 Cobbles used out of doors

been widely used as building materials. House fronts have been covered with row upon row of them, and cobbled roads still exist in the old narrow lanes of villages and towns in Britain and other countries. Alas, many are now being replaced by a smoother more practical, but to my taste, not anything like so attractive surface.

Cobbles are still a favoured way of giving variety to the flat planes of modern buildings; They are cemented into concrete tubs and walled enclosures, where once geraniums would have been planted, or decorate the forecourt of some multi-storied, plate glass and steel structure, or car park entrance.

If you would like to have cobbles interspersed among the paving slabs of a patio, or set in the wall of a garden room, or even have them loose in a sunken area, in this quantity you would have to obtain them from a dealer with a licence from the authority concerned, who has been given permission to collect a limited amount from a given beach, and who pays royalties for the privilege.

The pebbles themselves do not cost a lot by the hundredweight; transporting them is much the costlier item, unless you can arrange with the firm to collect them yourself. Some addresses of firms who supply cobbles are given at the end of the book.

MOUNTING MINERAL SPECIMENS

Beautiful specimens of mineral crystals, and fine pebbles too, can be mounted on a base or a plinth to show them to the best advantage, and to convert them into mineral and rock sculptures.

Transparent sheets of plastic, or solid blocks, such as Perspex clear plastic, are suited to this, and look elegant and modern. A cube or mount can be fabricated from Perspex, either by a firm which specialises in this work, or by buying the sheet Perspex or other plastic yourself, and the cement required, and making your own.

Perspex is a relatively expensive material, and fabricating mounts adds to the cost, but if you have a superb crystal specimen you might consider it worth your while. In this case you would need to send a sketch and measurements for an estimate to a firm which

FIG. 49 Fluorite specimen cemented on perspex block

fabricates plastic. Addresses are given at the end of this book where you can buy the material, or have the work done.

Polished or plastic lacquered wood mounts are also suitable, and you can probably use ingenuity to make up elegant stands from aluminium or brass bought from the local joinery and tools supplier.

The simplest and least expensive answer of any, is to pounce on any neat transparent plastic container which comes into the shops filled with sweets or stationary, or to use ready-made plastic jars of plain design from the kitchen departments of stores or shops. They could be trimmed to size with a saw and sandpaper, and make excellent mounts for minerals.

Appendix—Further Study

MUSEUMS

Institute of Geological Sciences,
Geological Museum,
Exhibition Road,
London, S.W.7

Natural History Museum (Mineralogy Department),
British Museum (Natural History),
Cromwell Road,
London, S.W.7

Many other museums in Britain have geology sections, and a full list can be seen in the subject index of *Museums and Galleries of Great Britain and Ireland*, an ABC Travel Guide published annually.

At the two museums above lectures with films and slides are given to the public starting on the days arranged at 3 p.m. A list of current lectures is available from each museum.

Those at the Geological Museum include lectures about rocks, volcanoes, identifying minerals and crystallography. At the Natural History Museum, many lectures are about the living world, but lectures are included about fossils, and shells, and other items of interest to students of rocks and minerals.

The Geological Museum has, in addition to displays of minerals and precious stones, complete ranges of rock specimens from every part of Britain, and relief geological maps on show.

EVENING AND DAY CLASSES

Most education authorities publish a list of local adult education classes in the late summer, and registration takes place in September. You are likely to find that classes in jewellery-making are available very widely; geology classes are fairly frequent, or lectures in the subject. Courses leading to qualifications are usually held at adult education colleges and polytechnics, and would be listed in their prospectuses.

A gemmology course is offered in London at Sir John Cass College of Art, and the Gemmological Association of Great Britain, Saint Dunstan's House, Carey Lane, London, EC2 offers correspondence courses, and holds examinations available to overseas students as well as students resident in Britain. The examinations may be held in many countries throughout the world.

The course covers two years and is divided into two parts, Preliminary and Diploma. It is not compulsory to sit the examination. A wide range of gemmological instruments, accessories and books are also obtainable from the society's address, and a catalogue is available.

A number of amateur geology and lapidary clubs and societies exist in most countries where the rock hobby is popular. In Britain, a list is published in the journal:

Gems (The British Lapidary Magazine)*
29, Ludgate Hill,
London, E.C.4

Australia
The Australian Lapidary Magazine,
Jay Kay Publications,
11, Robinson Street,
N.S. Wales.

Canada
Canadian Rockhound,
P.O. Box 194, Station A,
Vancouver 1,
British Columbia.

U.S.A.
Gems and Minerals,
P.O. Box 687,
Mentone,
California 92359.

FURTHER READING

Fossils, J. F. Kircaldy (Blandford Press, London).
Gem Cutting (A Lapidary's Manual), John Sinkankas (Van Nostrand).
Gemstones and Minerals, John Sinkankas (Van Nostrand).
Geology Made Simple, William H. Matthews, Advisory Editor and Eric A. Jarman, B.Sc. (W. H. Allen, London)
Minerals and Rocks in Colour, J. F. Kircaldy (Blandford Press, London).
Practical Gemmology, Robert Webster (NAG Press Ltd., London).
Rocks and Minerals, Zim and Shaffer (Hamlyn, London).
The Australian Gemhunter's Guide, K. J. Buchester (Ure Smith).

* *Gems* (The British Lapidary Magazine) also publishes a map, *Gemstone Sites of the British Isles*, which is revised and kept up to date. The magazine is also publishing a book, a guide to rock sites in Britain.

Other journals include:

The Gemmologist's Compendium, Robert Webster F.G.A. (NAG Press Ltd., London).
The Pebbles on the Beach, Clarence Ellis (Faber, London).

RECOMMENDED BOOKS FOR PEBBLE GARDENS

Gardens in Miniature, Anne Ashbury (Brockhampton Press).
House Plants, Joan Compton (Hamlyn, London).

PEBBLE ADDRESSES

Rock Shops and Lapidary Equipment

Ammonite Ltd.
Llandow Industrial Estate,
Cowbridge,
Glamorgan,
Wales.

M. L. Beach (Products) Ltd.,
41, Church Street,
Twickenham,
Middlesex.

Bryden Ltd.,
18, Wordsworth Road,
Colne,
Lancashire.

Craftorama (Wholesale) Co.,
14, Endell Street,
London, W.C.2.

Gemrocks Ltd.,
7/8, Holborn,
London, E.C.1.

Geobright,
28, Queens Road,
Brighton,
Sussex.

Hirsh Jacobson, Merchandising Co. Ltd.,
91, Marylebone High Street,
London, W.1.

Kernocraft Rocks and Gems, Ltd.,
44, Lemon Street,
Truro,
Cornwall.

Marbleshop,
Portsoy,
Banffshire,
Scotland.

P.M.R. Lapidary Equipment and Supplies,
Atholl Road,
Pitlochry, Perthshire,
Scotland.

J. W. Saundry,
12/13, Chapel Street,
Penzance,
Cornwall.

Selfridges Rock Shop,
Selfridges Ltd.,
Oxford Street,
London, W.1.

Staunton Gemstones and Craft Centre,
Staunton,
Nr. Coleford,
Gloucestershire.

Craft Supplies
Crafts Unlimited,
Macklin Street,
London, W.C.2.

Geological Maps and Books
Edward Stanford Ltd.,
12–14, Long Acre,
London, W.C.2.

Institute of Geological Sciences,
Geological Museum,
Exhibition Road,
South Kensington,
London, S.W.7.

Cutting Oils
Buck and Ryan Ltd.,
101, Tottenham Court Road,
London, W.1.

Burmah-Castrol Industrial Ltd.,
Special Products Division,
Marylebone Road,
London, N.W.1.
(Miracol 80).

N. Mole and Co. (Machine Tools) Ltd.,
7, Tolpits Lane,
Watford,
Hertfordshire.
(High grade soluble cutting oil supplied in quart amounts. Mail orders dealt with.)

Perspex and other Plastic
G. H. Bloore, Ltd.,
480, Honeypot Lane,
Stanmore,
Middlesex. And branches covering most of the United Kingdom.

Perspex and Fabrications
J. Watson (London) and Co., Ltd.,
Evelyn Road,
Acton Green,
London, W.4.

Pebbles, Cobbles and Other Minerals
British and Overseas Minerals Ltd.,
36, High Street,
Harrow-on-the-Hill,
Middlesex.

Index

Figures in bold type refer to rocks and minerals in charts